Cognitive Robotics and Adaptive Behaviors

Edited by Maki K. Habib

Published in London, United Kingdom

Cognitive Robotics and Adaptive Behaviors
http://dx.doi.org/10.5772/intechopen.95128
Edited by Maki K. Habib

Contributors
Miao Li, Xutian Deng, Thomas Tawiah, Wei Hong Chin, Naoyuki Kubota, Chu Kiong Loo, Axel Jantsch, Markus D. Kobelrausch, Diego Azevedo Leite, Alejandro Rafael Garcia Ramirez, Teddy Ordoñez Nuñez, Raimundo Celeste Ghizoni Teive

First published in London, United Kingdom, 2022 by IntechOpen
IntechOpen is the global imprint of INTECHOPEN LIMITED, registered in England and Wales, registration number: 11086078, 5 Princes Gate Court, London, SW7 2QJ, United Kingdom

British Library Cataloguing-in-Publication Data
A catalogue record for this book is available from the British Library

Additional hard and PDF copies can be obtained from orders@intechopen.com

Cognitive Robotics and Adaptive Behaviors
Edited by Maki K. Habib
p. cm.
Print ISBN 978-1-80355-387-0
Online ISBN 978-1-80355-388-7
eBook (PDF) ISBN 978-1-80355-389-4

Meet the editor

Maki K. Habib gained his Ph.D. in engineering science in intelligent and autonomous robots from the University of Tsukuba, Japan. He is a full professor of robotics and mechatronics in the Mechanical Engineering Department, School of Sciences and Engineering, The American University in Cairo, Egypt. His main areas of research interest are autonomous vehicles, human adaptive and friendly mechatronics, service robots and humanitarian demining, autonomous aerial vehicles and autonomous underwater vehicles, tele-cooperation, distributed teleoperation and collaborative control, flexible automation, ambient intelligence, biomimetic and biomedical robots, and Industry 4.0, including IoT, CPS, WSNs, and intelligent control.

Contents

Preface

Natural cognition systems, such as those of animals, humans, and many in nature, inspire the development of cognitive robots, which is an emerging interdisciplinary field in robotics. These robots represent forms of embodied cognition that focus on predictive capabilities, anticipate intended actions by perceiving their operational environments, and determine the necessary decisions and motor control. The cognitive robotics field describes robots that are continuously evolving and can achieve their goals by perceiving and interacting with their natural environment, recognizing and understanding events of interest, conducting adaptable planning, and anticipating the outcome of their actions and the actions of other entities sharing the same environment. These interactions enable the development of cognition capabilities through effective sensory-motor coordination. These robots use learning dynamics to exploit the full power of these interactions to deal with environment and task uncertainty and engage in continuous real-time reasoning.

This book provides up-to-date research development in the field of cognitive robotics. Topics covered include (but are not limited to) cognitive robotics, intelligent behaviors, systems intelligence, adaptive robotics, nature and bioinspiration, cognition architecture, cognitive modeling, knowledge representation, machine learning techniques, deep learning techniques, human-robot interaction, and evolutional robotics.

The six chapters contribute to the state-of-the-art and up-to-date knowledge on research advances in the field of cognition and robotics, introducing research at the interface between biology, sciences, engineering, and technology. With this book, we aim to develop a line of transformative research directions based on the adaptation of creative design and using intelligent methodologies, algorithms, and solutions. Tasks can be solved, and direct and indirect interaction with the task environment is developed by building evolving experiences through real-time learning. Cognitive robotics, AI, and machine learning allow researchers to think outside the box and open the way for new scientific challenges and developments.

Maki K. Habib
Mechanical Engineering Department,
The American University in Cairo,
New Cairo, Egypt

Chapter 1

The Neo-Mechanistic Model of Human Cognitive Computation and Its Major Challenges

Diego Azevedo Leite

Abstract

The neo-mechanistic theory of human cognition is currently one of the most accepted major theories in fields, such as cognitive science and cognitive neuroscience. This proposal offers an account of human cognitive computation, and it has been considered by its proponents as revolutionary and capable of integrating research concerning human cognition with new evidence provided by fields of biology and neuroscience. However, some complex cognitive capacities still present a challenge for explanations constructed by using this theoretical structure. In this chapter, I make a presentation of some of the central tenets of this framework and show in what dimensions it helps our understanding of human cognition concerning aspects of capacities, such as visual perception and memory consolidation. My central goal, however, is to show that to understand and explain some particular human cognitive capacities, such as self-consciousness and some conscious informal reasoning and decision making, the framework shows substantial limitations. I conclude the chapter by suggesting that to fully understand human cognition we will need much more than what the neo-mechanistic framework is actually able to provide.

Keywords: theoretical cognitive science, human cognitive computation, consciousness, informal reasoning, decision making and action

1. Introduction

A new intellectual movement in the field of cognitive science[1] has been developed, above all, in the last two decades of the current century, starting from debates that took place, mainly, in the philosophy of science at the end of the twentieth century. This movement has been described more broadly by many authors as a "new mechanistic philosophy" [4–7]. Strongly influenced by recent advances in computer science, neuroscience, and artificial intelligence, the theoretical framework developed by some

[1] I will use the term "cognitive science" in a *general sense* and a *specific sense*. In the general sense, the term will be treated as synonymous with the term "psychology" [1, 2]. In a specific sense, it will be treated as an attempt to build a science of cognition, integrating several different areas of knowledge, which took place in the 1970s in the USA [3].

IntechOpen

of the movement's most prominent authors offers a new physicalist (or materialist) and mechanistic view of human cognition[2] [9–21].

The theory formulated from the application of the neo-mechanistic philosophy to cognitive science and, specifically, to human cognition, can be called the *Mechanistic Theory of Human Cognition* (MTHC) [22]. This proposal is currently one of the most accepted major theories in fields, such as cognitive science and cognitive neuroscience, and it has been considered by its influential proponents as revolutionary and capable of integrating research concerning human cognition with new evidence provided by fields of biology and neuroscience.

One of the most central elements present in the framework of MTHC is a "model of human cognitive computation" [9–11, 13, 15], which is also part of the attempt made by several influential authors to provide some type of unification or integration for the field of cognitive science [9, 10, 23–25]. However, some complex cognitive capacities and some particular aspects of human cognition still present a challenge for explanations constructed by using this theoretical structure [22].

My central goal in this chapter, therefore, is to present an argument to show that human cognition cannot yet be completely understood and explained in terms of mechanistic computation and that this view indeed presents many substantial limitations.

To develop my argument, I present, firstly, some of the central elements of this neo-mechanistic framework and its application to cognitive science. Secondly, I present the mechanistic model of human cognitive computation, as it is currently framed, and, based on the specialized literature, I show in what dimensions it helps our understanding of some aspects of human cognitive capacities, such as visual perception and memory consolidation. Thirdly, I show that to understand and explain some human cognitive capacities, such as self-consciousness and conscious informal reasoning and decision making, the neo-mechanistic framework shows substantial limitations. I conclude the chapter by suggesting that the notion of human artificial cognitive computation can be useful for several projects, but to fully understand natural human cognition we will most certainly have to consider theories that go beyond the current neo-mechanistic model of human cognitive computation.

2. Mechanistic theory of human cognition

The contemporary movement of neo-mechanistic philosophy has been historically associated with ideas already present in the period of Ancient Philosophy. Philosophers, such as Democritus, Leucippus, Aristotle, Epicurus, and Lucretius [9, 14, 26], for example, have been mentioned in the specialized literature as precursors. Although there is no unity of thought regarding this philosophical tradition, these thinkers would arguably have launched, in Western philosophical thought, the first notions linked to mechanistic reflections. In other words, these philosophers would have proposed the general idea that many phenomena in nature must be explained through their basic components, their forms of movement, their properties, and their interactions since these phenomena are also composed of these basic elements.

In Modern Philosophy, the history of what might be called "mechanistic philosophy" is quite complex, given the many debates over definitions of the term and the

---

[2] I will use the term "cognition" as synonymous of the term "mind" for the sake of clarity and objectivity. For an important discussion concerning the term "cognition," cf. Akagi [8].

variety of positions that can be considered within a more general view of what the term means in this period. In any case, many authors consider that the movement of mechanistic philosophy in the seventeenth century is a reaction to Aristotelian natural philosophy and various natural philosophies of the Renaissance period [27]. The French philosopher René Descartes (1596–1650), for example, is considered one of the main figures who laid the foundations of modern mechanistic philosophy, especially with regard to explanations of biological natural phenomena [9, 27–31]. Des Chene [30] argues that Descartes united a mechanistic ontology, on the one hand, with a method of mechanistic explanation, on the other, applying these ideas to numerous biological phenomena, including the behavior of non-human animals and the human body.

Shortly thereafter, this reasoning would also be applied quite influentially to human beings and their mental capacities. One of the most prominent advocates of this view was the French philosopher and physician Julien Offray de La Mettrie (1709–1751), who published *Histoire Naturelle de L'âme* (Natural History of the Soul), in 1745, and *L' Homme Machine* (Man a Machine), in 1747, expanding Descartes' philosophy of biology to human beings [21]. It can be said, therefore, that modern mechanistic philosophy is fundamentally committed to the "machine analogy," that is, just as it occurs in a machine, all-natural processes can be explained in terms of their constituent components and the interaction between the activities they perform to produce their result [32]. This mechanistic framework was quite influential in many dimensions of many central issues and debates during the eighteenth and nineteenth centuries.

At the beginning of the twentieth century, the debate about the best explanation for the complex phenomenon of "life" was still quite strong [32]. The controversy was over whether or not this phenomenon could be explained in mechanistic terms. In this context, a very influential work was that of the German-born American physiologist and biologist Jacques Loeb (1859–1924), published in 1912, *The Mechanistic Conception of Life*. In this work, Loeb [33] indicates his interest in discussing the question of whether "life" (or all vital phenomena) could be explained in physicochemical terms. He sought to reduce "higher-level" biological phenomena to their more basic "low-level" components and thus ultimately place biology on the same level of scientific prestige and legitimacy as physics and chemistry [28].

In the second half of the twentieth century, philosophers of science sought to analyze, in a more precise way, this mechanistic explanatory strategy. One of the most influential analyzes is present in the work of the American philosopher Ernest Nagel (1901–1985), *The Structure of Science*, published in 1961. Chapter 12 of this work is entitled *Mechanistic explanation and organismic biology*. In it, Nagel [34] discusses the problem of explaining "life" and says that a mechanist is one who believes, as Jacques Loeb believed, that all vital processes can be explained in physicochemical terms. This work profoundly influenced the understanding of what a mechanistic scientific explanation was in the philosophy of science of the period.

It was also during this period that some philosophers of science working in the field of biology began the task of elaborating, in an even more robust and systematic way, notions related to mechanistic explanations in science – mainly in biology. Along these lines, some pioneering works were the following: Herbert Simon, *The Architecture of Complexity*, published in 1962; Stuart Kauffman, *Articulation of parts explanations in biology and the rational search for them*, published in 1970; and William Wimsatt, *Reductive explanation: a functional account*, published in 1976.

Within this line of philosophical thinking, the work of William Bechtel and Robert Richardson, *Discovering Complexity*, published in 1993, is normally considered in the specialized literature as being the first to elaborate mechanistic explanations of a more solid, detailed, and mature form. Moreover, in 1996, Stuart Glennan published the article *Mechanisms and the nature of causation*; in 1998, Paul Thagard published the article *Explaining disease: correlations, causes, and mechanisms*; in 2000, Peter Machamer, Lindley Darden, and Carl Craver published the article *Thinking about mechanisms*; and in 2002, Jim Woodward, published the article *What is a mechanism? A counterfactual account*. All these works were extremely important for the development of the new mechanistic movement in the philosophy of science, especially related to biology.

It is also important to point out that in the development of the neo-mechanist movement, at the end of the twentieth century, we can distinguish, more generally, two main trends [5]. One of them focuses more on metaphysical and ontological directions. Authors who work in this line seek, above all, to answer what mechanisms are as real things in the world. The other strand followed in the direction of a greater elaboration of the philosophy of science with epistemological and methodological discussions about scientific explanations, mainly in the area of biology. They seek to explain how something works and not make claims about the ultimate reality of things. These two strands of the new mechanism have been elaborated in an enormous specialized literature that covers several scientific and philosophical areas, dominating a great part of the central debates. Despite being two dimensions that can be separated in the debate, ontological and epistemological discussions are deeply related in many works, both directly and indirectly.

The neo-mechanistic philosophy began to be applied with greater emphasis to cognitive science since the decade of 1990 – with this application becoming stronger in the first decade of the twenty-first century – and it has been better elaborated since then until the present days in central works of very influential authors [9–15, 18–20, 35–43]. According to this view, human cognition, specifically, as well as biological cognition, in general, can be understood and explained through complex models of multilevel neurocognitive mechanisms. At these levels, there are causal processes related to cognitive information processing, cognitive representation, cognitive computing, as well as processes related to chemical and physical reactions that can be used to explain a given cognitive phenomenon. These are, in fact, autonomous processes of causation, which take place at all these different levels and are relevant to the explanation of the phenomenon of interest [44]. According to this theory of human cognition, namely, MTHC, all these causal levels and processes, although autonomous, can be related in a pluralistic mechanistic explanation, where the relevant scientific theories are integrated. As a result, MTHC includes not only a theory of human cognition but also a theory of the human neurocognitive relationship; that is, the theoretical framework suggests a possible solution to the problem of how we are to understand and explain the connection between human neural and cognitive phenomena, thus attempting to relate neuroscience and cognitive science.

The main objective of a mechanistic scientific explanation in scientific areas, such as biology, cognitive neuroscience, and cognitive science, is to identify the parts of a mechanism, its operations, its organization, and thus show how these elements constitute the system's relationship with the phenomenon that must be explained [9, 10, 45]. Particularly, in cognitive science, the central idea present in the theory is that human neurocognitive processes are a type of information processing performed by neural systems (mechanisms). These processes and the components that carry them out can be decomposed into subparts, and these subparts are decomposed again, as far as necessary for the understanding of the investigated phenomenon. After that, these components and activities have to be located in the brain as spatiotemporal parts of a complex multi-level neurobiological mechanism. As a result, there may be multiple levels of mechanistic composition in a human neurocognitive mechanism.

Another important feature of MTHC is that it was developed within a broad physicalist context that is present in a vast amount of work in contemporary cognitive science, philosophy of cognitive science, and philosophy of mind. In this physicalist context, the theory tries to combine central ideas present in traditional cognitive science with the main ideas present in certain fields of neuroscience that investigate human cognition. In this sense, some authors argue that this mechanistic physicalist framework can provide a consistent way to build a unified science of cognition and integrate cognitive science and neuroscience [23–25, 40].

Indeed, integrating and unifying, from a physicalist background, traditional cognitive science and traditional neuroscience to understand and investigate human cognition is an old dream held by many authors. Patricia Churchland, in 1986, calls for the unification of cognitive research and neural research in her book *Neurophilosophy: Toward a unified science of the mind-brain*. The aim of Churchland's book was to outline a general framework that would be suitable for the development of a unified theory of what she called "mind-brain," as well as to encourage the interaction between philosophy, psychology, and neuroscience [46].

It is possible to argue that MTHC was articulated with the objective of providing this integration and unification in a more precise theoretical way and within a clear physicalist background. The influential version of MTHC by William Bechtel is a clear example. He considers the human phenomenon "mind-brain" as "a set of mechanisms for controlling behavior" [9], and he explains that cognitive phenomena (e.g., perception, attention, memory, problem solving, and language) can be characterized as "information-processing mechanisms" [9]. Bechtel [9] states that scientific disciplines that aim to explain cognitive activities recognize that "in some way, these activities depend upon our brain." Or, to put it in another way: "Psychological phenomena are realized in brains comprised of neurons" [45]. This means that cognitive phenomena are physical and need to be explained in some physical (neural) way.

Craver and Tabery [47] describe the physicalist commitment quite clearly—"many mechanists opt for some form of explanatory anti-reductionism, emphasizing the importance of multilevel and upward-looking explanations, without rejecting the central ideas that motivate a broad physicalist world-picture." Therefore, in this approach, there is no space for any form of dualism, pluralism, or non-physicalism of any kind in relation to the ontology of human cognition. There is, indeed, a clear commitment to a form of ontological monism, namely, physicalism, that underlies the neo-mechanistic theory of human cognition.

Neo-mechanistic ideas about human cognitive phenomena are becoming increasingly dominant in fields related to theoretical cognitive science and cognitive neuroscience [48]. Consequently, the neo-mechanistic framework is often presented as one of the main theories, or the main theory, to explain human cognition in the twenty-first century.

3. Mechanistic model of human cognitive computation

Formulations of the idea that human cognition can be considered in computational terms can already arguably be found in the works of Thomas Hobbes

(1588–1679) and Gottfried Leibniz (1646–1716). However, it is in the first half of the twentieth century that new developments in this tradition made the thesis gain great strength [49]. Alan Turing (1912–1954), with his work on computation, made a solid mathematical contribution to advances in the attempt to build machines capable of thinking like humans. And with the development of the computer and the emergence of studies in computer science and artificial intelligence, there was an even greater push for the acceptance of these ideas in the period. Indeed, these were crucial factors in the development of cognitive psychology in the 1950s and cognitive science (in the specific sense) in the 1970s. In discussing the foundations of cognitive science, Gardner [3] states that "there is the faith that central to any understanding of the human mind is the electronic computer." Furthermore, according to him: "Involvement with computers, and belief in their relevance as a model of human thought, is pervasive in cognitive science" [3].

The first formulations of the philosophical foundations and the most central bases of the "computational theory of cognition" were presented, above all, in central works by Hilary Putnam (1926–2016) and Jerry Fodor (1935–2017). It is mainly based on works like these that the "classical model of cognitive computation" was formulated [49]. According to this proposal, the human mind is a computational system similar in important respects to a "Turing machine," which works through "Turing-style computations." In this view, cognitive processes, such as problem solving, decision making, and formal reasoning, are performed through computations similar to those of a Turing machine.

Another line of work, however, developed an alternative notion of cognitive computation. Inspired by research in the field of neurophysiology, some authors in the 1980s proposed that cognitive computation was something very different from Turing-style computation [50]. The correct format of cognitive computation for them was that of neural networks, in which, very briefly, data nodes are connected in a particular way so that when the network is activated through an input, it can provide an output. This framework became known as connectionism, and it has been developed in numerous works since then. Many cognitive models of different phenomena were built based on this view, such as object recognition, speech perception, and sentence comprehension.

The notion of "cognitive mechanistic computation" is part of this tradition, and it is especially related to the model of neural networks. Craver [10], for example, writes about the "computational properties of brain regions" and "computational properties of neural systems," without giving much detail about what exactly this means. In any case, it is clear that the supposed computation is much more related to concrete properties of neural systems than to abstract functional properties of psychological capacities considered in terms of Turing computation or something similar. Milkowski [11], in turn, presents a proposal that holds that neurocognitive processing occurs over states that contain information, but he does not elaborate much on the content and the semantic dimension of cognitive information or of putative cognitive computations.

Bechtel [9, 19] considers mental mechanisms as information-processing mechanisms that operate through neural representations and neural computations about vehicles and content. In his view, the "control theory of dynamical systems" shows how content is placed in this context. And Thagard [14, 15] thinks that mental mechanisms operate through computations that take place on representations at the cognitive level and computations that take place at the neural and molecular levels. In Thagard's work, there is also recourse to the "theory of dynamical systems" (as in Bechtel's); however, just in his version of the mechanistic theory, there is a definite number of mechanistic levels and extensive discussion about the "semantic pointers theory" of Chris Eliasmith.

Finally, there is the work of Piccinini [12, 13, 51], which is one of the most theoretically sophisticated and detailed among neo-mechanists regarding such issues. The author defends a mechanistic neurocomputational theory of human cognition. In his view, the human nervous system is a functional mechanism that produces computations through the activation of neurons, while the processing occurs in vehicles according to rules. Cognitive capacities are explained then by multilevel neurocognitive mechanisms that perform neural computations over neural representations. Besides, he thinks that neural computation (i.e., computations defined on the functionally relevant elements of neural activity) is not purely digital, as classically understood, nor purely analog, as alternatively understood; in his view, neural computation is *sui generis* – neither wholly digital nor wholly analog.

One does not need to enter so deep into these individual theories to see that they differ significantly. Craver mentions computations but does not offer an elaborated account. Thagard is the only one mentioning semantic pointers as central to the account. Milkowski and Piccinini attempt to avoid the problems with content, by means of focusing on formal properties. And Bechtel uses control theory to deal with the issue of content. As a result, it is not possible to derive from those accounts a single theory, as each author develops his own point of view with its significant particularities. There is, therefore, no theoretical substantial unity among these proponents.

However, one can try to find common aspects to evaluate at least the most basic and important tenets. To do that, an analysis of two cases where this mechanistic view on human cognitive computation can be applied will be helpful.

One of the best examples found in the specialized literature of a concrete application of this view to particular cognitive phenomena is related to memory, which, indeed, has been traditionally an object of study in the field of psychology [9, 10]. Functional analyses of the human memory capacity reveal the existence of many subcapacities, such as short-term memory, long term memory, phonological memory, visuospatial memory, semantic memory, episodic memory, and memory consolidation. In mechanistic terms, one of the best-understood phenomena in this memory system is memory consolidation. Roughly put, this is the phenomenon of transforming short-term memories (which are liable and easy to disrupt) into long-term memories, which are robust and enduring, when consolidation takes place and permits the organism to remember important events for a longer period of time and modify its behavior accordingly [52]. To explain this phenomenon, all the relevant regions in the brain responsible for the functions that compose the neuro-cognitive mechanism of memory consolidation, including all relevant mechanistic levels of decomposition, must be identified, that is, all the particular component parts and component operations of the whole mechanism must be determined, as shown on **Figure 1**. Finally, the causal processes and causal interactions within the mechanism functions need also to be understood, that is, the general organization of the mechanism.

The explanation starts at the highest level of the whole mechanism. At this level, it is necessary to correctly identify all the large neural network that is responsible for memory consolidation. Secondly, it must be established whether this large neural system is indeed all that is relevant for the explanation of the phenomenon. The mechanistic explanation at this level also needs to clarify how the neural network process information about new memory episodes through *computational operations*

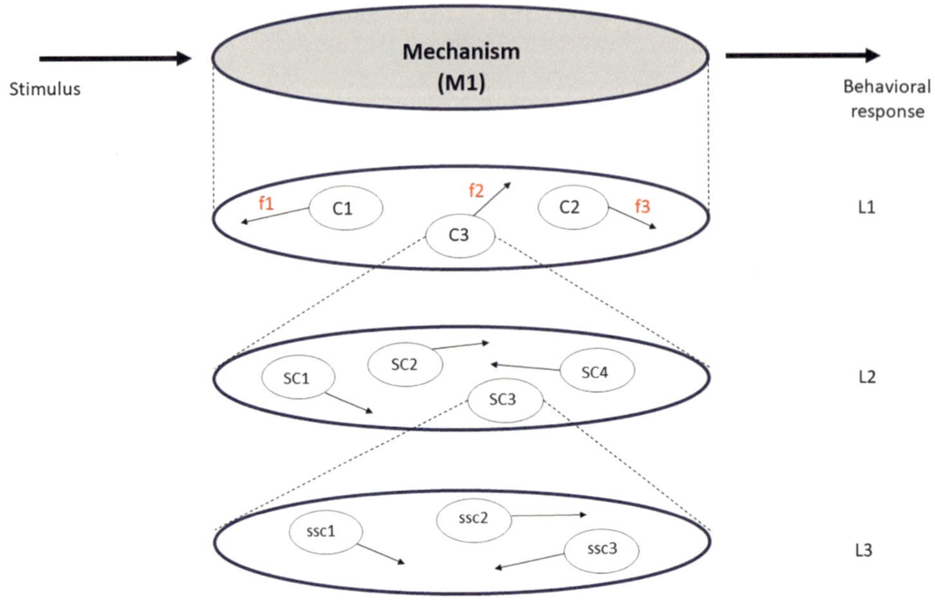

Figure 1.
An example of a simple model of a neuro-cognitive biological mechanism (M1). In this model, M1 is composed, at the level L1, by its component parts C1, C2, and C3, which perform the functions (or activities) f1, f2, and f3. The component parts can be decomposed into smaller components, as it happens with C3, which is composed, at level L2, of the sub-components SC1, SC2, SC3, and SC4. The component SC3 can be further decomposed, at level L3, into its subcomponents ssc1, ssc2, and ssc3.

and how these processes produce and affect, for instance, the different degrees of consolidation that characterize the memories under investigation.

Once this has been clarified, the explanation turns to the second level of description in which the large neural system is decomposed into particular sub-neural systems localized in more specific regions. Here the goal is to understand the information processing and computational operations (e.g., spiking patterns in populations of neurons) of these smaller neural networks and how they contribute to the performance of the whole mechanism composed of such neural nets.

Moreover, a further stage of decomposition must be reached that concerns the processes underlying memory at an intercellular level. The explanation at this particular level aims at describing the components of a particular neural network and at understanding how a small number of neurons operate (e.g., how they depolarize and fire in the process of propagation of action potentials, or how they are responsible for synaptic processes, neurotransmitters being released, and so on). Here it is possible to measure spiking rates of neurons, or spiking frequency and record neural activity in general.

Finally, the explanation can go even to another lower mechanistic level—the intracellular and molecular level. At this level, the description is in terms of the activity of relevant proteins, molecules, and ions. As one can see, this kind of explanation "exhibits a progression from the behavioral-level characterization of memory consolidation to the identification of important components in the process at progressively lower levels." [52]. All levels are equally important to achieve the complete multilevel mechanistic explanation of the particular phenomenon in the end.

Another example is related to human visual perception [9, 13, 40], which is roughly understood as the capacity to acquire and process visual information from objects and events in the environment. In the biological mechanism related to human visual perception, the occipital lobe is central, since many studies on humans show deficits in visual processing due to damage in the occipital lobe. The mechanism also includes a projection of the optic tract going from the eye, passing by the lateral geniculate nucleus (LGN), which is an area of the thalamus, and achieving the occipital lobe. Besides, it includes the eyes, optic nerves, and other brain areas responsible for visual perception. All these areas can be decomposed in working components and their operations, and each decomposition is considered to be a lower level in the entire constitution of the mechanism. The occipital lobe, for instance, can be itself decomposed in areas responsible for particular visual functions, such as the striate cortex, also known as Brodmann area 17, or V1 (primary visual cortex, or visual area 1).

The same procedure can be done for all the other areas in the brain that are also part of the mechanism responsible for visual perception; for instance, V2, V3, V4, and V5/MT. It is necessary to identify also the cells (including visual receptor cells in the retina of the eye, such as cones and rods), networks of cells, or larger neural systems in these areas that are responsible for *information processing and computation*, for example, about light and dark spots, bars of light (edges), size, shape, color, depth, location, and motion of objects in the visual field. The mechanism also includes the pathways and channels through which the information is transmitted and the information about intercellular, intracellular, and molecular processes.

As one can observe by looking at these two examples, the notion of "computation" in the mechanistic framework stands for some causal interactions within the nervous system and this is how different brain regions "compute" different information. Each brain region "stands for" some kind of particular information—related to perception, sensation, memory, language, reasoning, emotion, etc. The substantial problems with such an account of human cognition will be analyzed in what follows.

4. Major challenges to the model

A great deal of criticism has arisen in the specialized literature concerning the notion of human cognitive computation. It is nearly impossible to review all of the works, but I will make some considerations of some of the most influential critics.

Fodor [53–55], for instance, claims that many mental representations (e.g., beliefs) and mental processes (e.g., abductive reasoning) are sensitive to global properties (i.e., properties that beliefs, for instance, have so that they are determined by a set of other beliefs which they are members of). For example, a belief about a tennis racket being broken may complicate the plan of playing tennis on the weekend, but not the plan of playing soccer. This means that a mental representation, such as an intention to play tennis, will depend on the context at the moment—whether there is a racket available for the game or not. Fodor argues, though, that classical symbolic computing models are only sensitive to local properties, and neural network models cannot handle this feature of human cognition.

Dreyfus [56], in turn, claims that much human knowledge cannot be captured by symbolic manipulation and formal rules, since this knowledge is constructed through direct contact and practicing in the world. Nagel [57] brings attention to the problem of phenomenal consciousness—roughly, the issue of what it feels like to experience something subjectively. Following this line of thinking, we can also say that a computer cannot know (if it can know anything) what it feels like to taste the flavor of chocolate. It has no idea of what it is like to eat chocolate, something that is quite basic for any child that does it. More than that, computers do not feel pain or pleasure, which is quite basic for human beings. Furthermore, Searle [58] brings attention to the difficulties related to intentionality, understanding, and meaning, with his famous "Chinese room argument." And, additionally, Putnam [59] develops the idea that mental states cannot be identified with computational states, consequently arguing vigorously against computational reductionism[3].

The case of Bruner's critics is also very interesting. One of the names most frequently mentioned in influential works of historical reconstruction of the events and studies that contributed to the beginning and development of the cognitive movement in psychology is the American psychologist Jerome Bruner (1915–2016) [1–3, 60, 61]. He is recognized for having founded, together with George Miller (1920–2012), the Center for Cognitive Studies at Harvard University, in 1960. In addition, Bruner published, together with colleagues, in 1956, *A Study of Thinking*, in which he dealt, in a systematic way, with the formation of concepts under a cognitive perspective, which gave great impetus to the movement. In his various works, Bruner has contributed to scientific knowledge on various topics of psychology, such as perception, language, learning, and cognitive development [62].

One of the most interesting points in Bruner's work, however, is his strong criticism of the very cognitive movement he helped to develop. He has presented this criticism in key works, such as *Acts of Meaning*, published in 1990, and *The Culture of Education*, published in 1996. Examination of these works can thus show what an author with a rigorous background in scientific psychology, a high degree of theoretical sophistication, and extensive research in the field observed that was wrong with the development of cognitivism.

In *Acts of Meaning*, Bruner [63] states that the original idea of the cognitivist movement of the 1950s was, in fact, to establish "meaning" as a central concept of psychology. However, in Bruner's view, this original impulse was distorted by a reductionist emphasis, adopted by a dominant trend of the movement that defended computationalism. The emphasis was given to "information," "processing of information," and "computability;" and not to meaning and to "meaning construction" [63]. As a result of this approach, concepts central to traditional inquiry in scientific psychology have been distorted, eliminated, or obscured, such as the concepts of "intentional states" (believing, desiring, intending, understanding a meaning) and the concept of "agency," that is, the conduct of human action under the influence of intentional states [63].

However, in Bruner's view, this is not the way forward. In *The Culture of Education*, Bruner [64] says that, since the cognitive revolution, there have been two quite different conceptions of how the human mind works—the first establishes the hypothesis that the human mind works as a computational system; the second proposes the hypothesis that the human mind is constituted and realized in the use of human culture. Bruner claims that his version of cognitivism is not based on reductionist computationalism, but rather on what he called culturalism. He claims that his intention is really to develop a theory of the human mind alternative to computationalism

[3] Of course, these arguments are still being strongly debated currently, and there are many attempts to answer these concerns. If the answers are satisfying or not, it is something that cannot be settled here. However, in any case, these arguments taken together provide a very compelling case against the idea that all human cognition can be understood and explained in computational terms.

and that his theory focuses exclusively on "how human beings in cultural communities create and transform meanings" [64].

One of the major problems pointed out by Bruner in the computationalist approach is that the production of meaning is often extremely complex, sensitive to the context, and involves the difficulty of clear and precise understanding [64]. This is not the same as establishing computational procedures for the processing of input and output information to the system, whether this is computational processing in digital format or the form of neural networks. For Bruner, meaning making is not merely information processing; it is something more profound and more complex. Culture, in his view, has a fundamental role in human life and it is only through it and in it that certain processes and mental structures are formed and used.

The human being, in Bruner's view, was able to develop a way of life in which reality is represented by a symbolism shared by members of a cultural community, and human life is organized and built from this symbolism that is conserved, elaborated, and transmitted through successive generations [64]. Although meaning is in the mind and is produced by it, it also has its origins in culture and has its importance within the culture in which it was generated. And for the production of meanings, the human mind creates and makes use of symbolic cultural systems. Thus, in this view, thinking and learning are always situated in a cultural context [64]. Computer systems, however, are not capable of producing meanings. They only deal with a certain set of formalized and operationalized meanings, but they do not make interpretations of human and cultural phenomena.

Furthermore, there is no very clear reason to suppose that processes and relationships between all mental phenomena are literally computational in nature, nor that all mental representations have this same character. The application of the concept of computation to these phenomena investigated in the tradition of psychological research is based only on a working hypothesis present in a certain particular theoretical system. Nevertheless, there is as yet no concrete proof that all human cognition works according to a type of computational processing x, y, or z. In fact, finding out what kind of computational processing is related to the human mind has become an extremely debated issue internally by adherents of any computational model of human cognition [49]. It is no accident that comprehensive theoretical systems were developed precisely with the intention of questioning the computational model of cognition.

Now, to illustrate more concretely some of the difficulties mentioned with the notion of human cognitive computation, let us consider some cases involving conscious complex informal reasoning and conscious complex decision making where explanations for human behavior might be required [22].

Consider, firstly, a case where a person is dissatisfied with her marriage and is thinking about getting a divorce. To make such a decision, she has been consciously reflecting for months on the current state of the marriage, her beliefs about the relationship, her emotions about her partner, her desires and expectations in life, the beliefs of her family and closest friends about the issue and what are the reasons to take action in this regard. After thinking carefully for a very long time, being aware that she really does not feel comfortable and happy at all, she decides to go for a divorce.

Consider also a second example. A person needs to decide which candidate she will vote for as president of her country. To make this decision, she needs to use her conscious informal reasoning ability. Thus, she reflects on the arguments put forward by politicians running for the election, the arguments put forward by commentators, scientists, and political analysts, as well as journalists writing on the subject, and the arguments of friends and family she finds relevant and credible. After three months of thinking, she has not decided yet but is rather still in doubt concerning her vote in the major candidates A and B. When someone asks her which candidate she is going to vote, she says: "I still don't know." Then, some surprising news arises in a serious newspaper with charges of corruption against candidate A, and she is a frequent reader of this newspaper, so she becomes immediately aware of this. Upon reflection on the matter and related issues, she takes the new information seriously and she finally decides that voting for candidate B is the best option. The major reason is that there is no charge whatsoever of corruption against him. When she is asked now which candidate she is going to vote for, she answers immediately: "candidate B." After she made up her mind, she finally goes to the appropriate place on the proper day and time to cast her vote.

A third example is the case of a college student who suffers from difficulties related to his excessive anxiety. Through a general psychological assessment, it can be seen that the factors related to student anxiety are financial difficulties, difficulties in family life where physical and psychological violence occurs, difficulties in finding leisure time to relax and have fun (since they need to work and study at the same time) and difficulties with excessive concerns about the uncertain future, as he believes that it will not be easy to find a job when he graduates. All of these factors seem to contribute to generate in the student's mind distorted and dysfunctional negative thoughts about himself and his life, and it seems very plausible that these distorted thoughts are strongly associated with his excessive anxiety. This interpretation is, indeed, supported by numerous works in the specialized literature in clinical psychology. Thus, we observe that the most relevant causal factors to explain this psychological phenomenon are not merely computational, but psychological, social, and environmental.

Psychological scientific explanations, in these cases, need considerations that go beyond the investigation of computations being performed in nervous systems or even in any abstract functional system. What explains the psychological phenomenon of belief formation and decision making in the first example and the excessive anxiety in the third example is the meaning formation and interaction of beliefs, desires, and intentions to act (according to logical rules, practical rules, and interpretation of reality), which are strongly affected by emotions, physical environment, and social factors.

In the second example, evidently, an informative explanation would have to mention an important causal factor—the event of the corruption charges against candidate A, appearing in a serious newspaper. Moreover, the explanation would have to mention that the person becomes aware of this event, accepts it as reliable, accepts the charges as true and accurate, and now this content is present in one or some of her beliefs. In possession of this content, she can rationally justify herself when engaging in discussions about the topic with family, friends, and other people, providing reasons for her related beliefs and her related behaviors. Thus, the influence of the event on her is external and affects the internal logic and content of her systems of beliefs, emotions, desires, and intentions. This explanation involves then particular properties of human cognitive systems, present for instance in belief and intention systems. These properties are clearly different than those involved in merely describing supposed automatic computational activities in her neural networks or describing what is happening in terms of physical and chemical neural processes. The explanation for this phenomenon of belief formation, therefore, would also have to account for how this new information could change a particular belief given her system of beliefs about the topic.

In the examples above, there are cognitive processes that often necessitate consciousness and complex informal reasoning about belief systems that are often linked to particular perceptions, sensations, emotions, desires, intentions, attitudes, as well as related to each other and the external environment. Some of these beliefs have great value, such as some moral beliefs, which makes this whole dynamic even more complex. In these cases, blind computation might even occur at some level, but what is most relevant are environmental, social, cultural, historical, and psychological factors (such as beliefs, emotions, desires, and intentions) that acquire meaning in a given cognitive system.

The relevant explanation of the actions in such cases is made through considerations—(1) about the creation and alteration of the content of perceptions, beliefs, sensations, emotions, maxims, wills, desires, intentions, etc.; (2) about their internal relationships; and (3) about their external relationships with the physical, social, historical, and cultural context. Rigorous empirical scientific research can aid in discovering strong and systematic (stable) regularities in human behavior explained in such terms without the need for the notion of computation. Statistical tools and analysis, through the mathematical application, can bring greater objectivity, avoiding both an extremely subjective and confusing vocabulary, as well as unproductive speculation and mere common sense.

Moreover, self-consciousness here is crucial, since we humans have the ability to *evaluate* our own beliefs, not just to be aware that we have them. If we can access some beliefs as belonging to our cognitive belief system, we can evaluate whether they are true or false, precise or imprecise, how they are related to our emotions and sensations and we can decide if we want to keep them or not. The complex social dynamics are also crucial, since our systems of beliefs are constantly interacting with the beliefs of others during our lifetime and this interaction has a major influence on the formation and modification of our belief system, emotional system, and volitional system.

Therefore, human beings have the ability to form original belief systems and relate them according to logical and interpretative rules, building arguments to support their point of view, which often influences their behavior. Human beings are also able to think about different types of relevant information for months or years to make an important and complex decision. To make a difficult decision, a human being can take into account information related to plans for the very distant future, in which many scenarios are considered. A human may wonder what happened in the very distant past, or what might have happened, even if he or she knew what really happened. And complex informal reasoning and complex decision making are things that humans do naturally and often in their daily lives.

Thus, in cognitive science, it is necessary to deal with extremely complex phenomena, given that human beings show great differences when compared to other animals in nature. Human beings have a cumulative, complex, dynamic, and elaborate culture that is passed on through generations. Humans are also involved in understanding and writing their own history. They have natural languages with enormous, complex, and refined expressive power and sophisticated grammar. Human beings practice and appreciate art, such as literature, painting, cinema, and music. They engage in purely formal or very abstract thoughts when they do mathematics, logic, and engage in certain religious thoughts. They create legal laws for their societies and think about morality, building moral systems. They build artificial intelligence machines that are able to learn with a certain level of autonomy and are able to explore other planets. Furthermore, humans are involved in politics, science, and philosophy.

Computers, by contrast, so far, do not form beliefs on their own, they do not have the capacity to evaluate and improve them by themselves, and they do not interact in the social environment neither using natural language with a huge degree of sophistication as humans do nor engaging in social and cultural practices. If we look at the problem from a very concrete and objective point of view, we observe that even the most advanced computer systems, the most advanced robots, and the most advanced artificial neural and cognitive architectures today are still very far from behaving like human beings in relation to language and actions that involve consciousness and informal rationality. Humans are capable of playing chess, cooking pizza, making coffee, having a conversation about politics, creating a new song on a guitar, and playing tennis on the same day. No computational artificial system is currently capable of this generality in cognition. So, as a matter of current fact, computational artificial cognition cannot be used to fully explain the major capacities of human cognition and intelligence.

It is no surprise, then, that mechanistic accounts of psychological capacities usually suggest only *where* the putative computations are probably taking place in the idealized standard human brain (as we can see in the examples presented in the previous section), not *what* exactly are these computations and how they can be related to the internal subjective experience of a person (like the content of a strong belief, for instance, that can normally be accessed and become conscious).

Difficulties with the notion of cognitive computation are recognized by influential neo-mechanists themselves. Milkowski [21], for instance, concludes his work by admitting that we "still don't know how to model consciousness mechanistically." Additionally, there are several alternative models of cognitive computation in cognitive science nowadays—syntactic computation; algorithmic computation; causal computation; and semantic computation [65]. None of the models has gained significant prominence over the others concerning the understanding and explanation of human cognition. Finally, there is strong criticism even of the attempt by neo-mechanists to propose that good computational explanations in cognitive science must be also mechanistic explanations [66, 67].

Therefore, if we think about the issue from the point of view of current facts, we need to recognize that the neo-mechanistic proposal for human cognition is still far from being able to be considered the best or most plausible understanding and explanation of human cognition. It is just one view among many.

5. Conclusion

The mechanistic framework has been offering significant contributions to the field of cognitive science, on the one hand. One of its best contributions is the promotion of debates on the issue of human cognitive computation. In this sense, there is a search for a better understanding of what this notion actually means. All this effort is very worthwhile and welcome. More generally, the theoretical debate about fundamental questions in cognitive science promoted by new mechanists is also very important, as well as their effort to clarify what a "biological mechanism" and a "cognitive mechanism" are and what a "mechanistic explanation" in cognitive science is. Furthermore, another contribution of the new mechanistic philosophy is to encourage historical research and current debate, in cognitive science and beyond, about the relationship among "mechanism," "materialism," "reductionism" and "computationalism", so that these concepts are not confused and that the positions adopted by the authors, as well as the different dimensions of the debate, are appreciated in a fair and correct way. Finally, the new mechanistic philosophy applied to cognitive science is also contributing to the important debate concerning the unity, integration, and plurality in the field.

On the other hand, however, many of the current promises of the new mechanism for cognitive science are quite difficult to fulfill. Firstly, neo-mechanistic philosophy is a philosophy of science built primarily from examples from the biological sciences and neuroscience that is serving as the basis for building a philosophy of the science of mind. We live in a period in which neuroscience and artificial intelligence research have gained great prestige and recognition. A great deal of economic investment has been made in these areas and this is very attractive. In part, this also influences "the new wave of mechanism," and the necessity of some authors to expand the framework. However, numerous particularities related to psychology and human cognition are being neglected in this theoretical structure, as I tried to show.

Secondly, there is considerable disagreement among leading neo-mechanists over the most plausible formulation of MTHC regarding fundamental issues, such as the idea of human cognitive computation. Thus, there is a considerable difficulty related to the internal articulation and unification of the theory. Furthermore, many alternative major theories, and the research programs based on them, strongly threaten the neo-mechanistic framework in current cognitive science, since they are also seeking predominance in the field, or just for having more space and recognition.

Given this, we can conclude that the mechanistic model of human cognitive computation cannot provide substantial theoretical or explanatory unification or integration to the field of cognitive science today, since there is no unification between the proponents themselves. Moreover, their different proposals are often unclear on many important aspects concerning traditional problems of intentionality, consciousness, and self-consciousness. The accounts are sometimes internally not well-articulated; and, externally, there is serious criticism of them, with countless debates and controversies on several fundamental questions. In addition, there are several alternative models competing for predominance on this particular issue. And it is yet by no means clear whether the explanatory power of any of them is greater than the explanatory power of the others.

This analysis shows, therefore, that the neo-mechanistic proposal concerning human cognitive computation has serious weaknesses. But the problem is not to use the idea of cognitive computing to advance models of biological and artificial cognitive architectures, since many human cognitive abilities can already be simulated. Indeed, it is very interesting to see that our science has advanced to the point where a computer can win against the best chess and go game players in the world. In fact, advancements within computational artificial systems and robotics could well be applied to improve our educational and health systems. For example, inspired by scientific developments in the field of cognitive science, artificial cognitive systems could possibly be developed to help children with the learning process of mathematics, natural language, or history at schools, or even at the university level. Artificial systems could possibly be developed to help people with excessive anxiety symptoms, as well. This could be extremely worthwhile. Moreover, better and more advanced artificial cognitive systems and robotic systems can contribute to improving theories of human cognition, as much as better and more correct theories of human cognition can help in faster advancements of cognitive artificial systems and robotic systems. But there is good reason to keep these efforts separated and to consider human cognition as a very complex and particular phenomenon in nature.

The problem arises only with the untenable suggestion that we already have, or that we are very close to getting, the complete and definitive understanding and explanation of all the major capacities of human cognition in computational terms. This, yes, is a mistake.

Conflict of interest

The author declares no conflict of interest.

Author details

Diego Azevedo Leite
Federal University of Alfenas, Alfenas, Minas Gerais, Brazil

*Address all correspondence to: diego.azevedo@unifal-mg.edu.br

References

[1] Leahey TH. A History of Psychology: From Antiquity to Modernity. 8th ed. New York: Routledge; 2018

[2] Mandler G. A History of Modern Experimental Psychology: From James and Wundt to Cognitive Science. Cambridge, MA: MIT Press; 2007

[3] Gardner H. The Mind's New Science: A History of the Cognitive Revolution. New York: Basic Books; 1985

[4] Glennan S. The New Mechanical Philosophy. Oxford: Oxford University Press; 2017

[5] Glennan S, Illari P. Introduction: Mechanisms and mechanical philosophies. In: Glennan S, Illari P, editors. The Routledge Handbook of Mechanisms and Mechanical Philosophy. New York: Routledge; 2018. pp. 1-10

[6] Krickel B. The Mechanical World. Cham: Springer; 2018 Available from: http://link.springer.com/10.1007/978-3-030-03629-4

[7] Glennan S, Illari P, Weber E. Six Theses on Mechanisms and Mechanistic Science. Journal for General Philosophy of Science. 2021. DOI: 10.1007/s10838-021-09587-x

[8] Akagi M. Rethinking the problem of cognition. Synthese. 2018;**195**:3547-3570

[9] Bechtel W. Mental Mechanisms: Philosophical Perspectives on Cognitive Neuroscience. New York: Routledge; 2008

[10] Craver CF. Exlaining the Brain: Mechanisms and the Mosaic Unity of Neuroscience. Oxford: Oxford University Press; 2007

[11] Milkowski M. Explaining the Computational Mind. Cambridge, MA: MIT Press; 2013

[12] Piccinini G. Physical Computation: A Mechanistic Account. Oxford: Oxford University Press; 2015

[13] Piccinini G. Neurocognitive Mechanisms: Explaining Biological Cognition. Oxford: Oxford University Press; 2020

[14] Thagard P. Hot Thought: Mechanisms and Applications of Emotional Cognition. Cambridge, MA: MIT Press; 2006

[15] Thagard P. Brain-Mind: From Neurons to Consciousness and Creativity (Treatise on Mind and Society). Oxford: Oxford University Press; 2019

[16] Bechtel W, Abrahamsen A. Explanation: A mechanist alternative. Studies in History and Philosophy of Biological and Biomedical Sciences. 2005;**36**(2):421-441

[17] Machamer P, Darden L, Craver CF. Thinking about Mechanisms. Philosophy in Science. 2000;**67**(1):1-25

[18] Bechtel W. Resituating cognitive mechanisms within heterarchical networks controlling physiology and behavior. Theory & Psychology. 2019;**29**(5):620-639

[19] Bechtel W. Investigating neural representations: The tale of place cells. Synthese. 2016;**193**(5):1287-1321. DOI: 10.1007/s11229-014-0480-8

[20] Bechtel W. Looking down, around, and up: Mechanistic explanation in psychology. Philosophical Psychology. 2009;**22**(5):543-564

[21] Milkowski M. Mechanisms and the mental. In: The Routledge Handbook of Mechanisms and Mechanical Philosophy. New York: Routledge; 2018. pp. 74-88

[22] Leite DA. The Twenty-First Century Mechanistic Theory of Human Cognition: A critical analysis. Cham: Springer; 2021

[23] Milkowski M. Integrating cognitive (neuro)science using mechanisms. Avante. 2016;**6**(2):45-67

[24] Milkowski M, Hohol M, Nowakowski P. Mechanisms in psychology: The road towards unity? Theory & Psychology. 2019;**29**(5):567-578. DOI: 10.1177/0959354319875218

[25] Piccinini G, Craver CF. Integrating psychology and neuroscience: Functional analyses as mechanism sketches. Synthese. 2011;**183**(3):283-311

[26] Popa T. Mechanisms: Ancient sources. In: Glennan S, Illari P, editors. The Routledge Handbook of Mechanisms and Mechanical Philosophy. New York: Routledge; 2018. pp. 13-25

[27] Roux S. From the mechanical philosophy to early modern mechanisms. In: Glennan S, Illari P, editors. The Routledge Handbook of Mechanisms and Mechanical Philosophy. New York: Routledge; 2018. pp. 26-45

[28] Allen GE. Mechanism, organicism, and vitalism. In: Glennan S, Illari P, editors. The Routledge Handbook of Mechanisms and Mechanical Philosophy. New York: Routledge; 2018. pp. 59-73

[29] Boas M. The establishment of the mechanical philosophy. Osiris. 1952;**10**(1952):412-541

[30] Des CD. Mechanisms of life in the seventeenth century: Borelli, Perrault, Régis. Studies in History and Philosophy of Biological and Biomedical Sciences. 2005;**36**(2):245-260

[31] Gunderson K. Descartes, La Mettrie, Language, and Machines. Philosophy. 1964;**XXXIX**(149):193-222

[32] Allen GE. Mechanism, vitalism and organicism in late nineteenth and twentieth-century biology: The importance of historical context. Studies in History and Philosophy of Biological and Biomedical Sciences. 2005;**36**(2):261-283

[33] Loeb J. The Mechanistic Conception of Life. Chicago: The University of Chicago Press; 1912

[34] Nagel E. The Structure of Science: Problems in the Logic of Scientific Explanation. New York: Harcourt, Brace & World, Inc.; 1961

[35] Craver CF. Interlevel experiments and multilevel mechanisms in the neuroscience of memory. Philosophy in Science. 2002;**69**(S3):S83-S97

[36] Bechtel W. Multiple levels of inquiry in cognitive science. Psychological Research. 1990;**52**(2-3):271-281

[37] Bechtel W. Levels of description and explanation in cognitive science. Minds and Machines. 1994;**4**(1):1-25

[38] Bechtel W. Constructing a philosophy of science of cognitive science. Topics in Cognitive Science. 2009;**1**(3):548-569

[39] Bechtel W. How can philosophy be a true cognitive science discipline? Topics in Cognitive Science. 2010;**2**(3):357-366

[40] Boone W, Piccinini G. The cognitive neuroscience revolution. Synthese. 2016;**193**(5):1509-1534

[41] Bechtel W. Decomposing the mind-brain: A long-term pursuit. Brain and Mind. 2002;**3**(2):229-242

[42] Craver CF. The making of a memory mechanism. Journal of the History of Biology. 2003;**36**(1):153-195

[43] Craver CF. Beyond reduction: Mechanisms, multifield integration and the unity of neuroscience. Studies in History and Philosophy of Biological and Biomedical Sciences. 2005;**36**(2):373-395

[44] Craver CF, Bechtel W. Top-down causation without top-down causes. Biology and Philosophy. 2007;**22**(4):547-563

[45] Bechtel W, Wright CD. What is psychological explanation? In: Symons J, Calvo P, editors. The Routledge Companion to Philosophy of Psychology. New York: Routledge; 2009. pp. 113-130

[46] Churchland PS. Neurophilosophy: Toward a Unified Science of the Mind-Brain. Cambridge, MA: MIT Press; 1986

[47] Craver C, Tabery J. Mechanisms in science. In: Zalta EN, editor. The Stanford Encyclopedia of Philosophy. Summer 2019 Edition. 2019. Available from: https://plato.stanford.edu/archives/sum2019/entries/science-mechanisms/

[48] Samuels R, Margolis E, Stich SP. Introduction: Philosophy and Cognitive Science. In: Margolis E, Samuels R, Stich SP, editors. The Oxford Handbook of Philosophy of Cognitive Science. Oxford: Oxford University Press; 2012. pp. 1-18

[49] Rescorla M. The computational theory of mind. In: Zalta EN, editor. The Stanford Encyclopedia of Philosophy. Fall 2020 Edition. 2020. Available from: https://plato.stanford.edu/archives/fall2020/entries/computational-mind/

[50] Buckner C, Garson J. Connectionism. In: Zalta EN, editor. The Stanford Encyclopedia of Philosophy. Fall 2019 Edition. 2019. Available from: https://plato.stanford.edu/archives/fall2019/entries/connectionism/

[51] Piccinini G. Computing mechanisms. Philosophy in Science. 2007;**74**(4):501-526

[52] Bechtel W. Molecules, systems, and behavior: Another view of memory consolidation. In: Bickle J, editor. The Oxford Handbook of Philosophy and Neuroscience. Oxford: Oxford University Press; 2009. pp. 13-40

[53] Fodor J. The Mind Doesn't Work That Way: The Scope and Limits of Computational Psychology. Cambridge, MA: MIT Press; 2000

[54] Fodor J. Reply to Steven Pinker "So how does the mind work?". Mind & Language. 2005;**20**(1):25-32

[55] Fodor J. LOT 2: The Language of Thought Revisited. Oxford: Oxford University Press; 2008

[56] Dreyfus HL. What Computers Can't Do: A critique of artificial reason. New York: Harper & Row, Publishers; 1972

[57] Nagel T. What is it like to be a bat? Philosophical Review. 1974;**83**(4):435-450

[58] Searle JR. Minds , brains , and programs. The Behavioral and Brain Sciences. 1980;**3**:417-457

[59] Putnam H. Representation and Reality. Cambridge, MA: MIT Press; 1988

[60] Bechtel W, Abrahamsen A, Graham G. The life of cognitive science.

In: Bechtel W, Graham G, editors. A Companion to Cognitive Science. Oxford: Blackwell Publishing; 1998. pp. 1-104

[61] Miller GA. The cognitive revolution: A historical perspective. Trends in Cognitive Sciences. 2003;7(3):141-144

[62] Greenfield PM. Jerome Bruner (1915-2016). Nature. 2016;**535**:232

[63] Bruner J. Acts of Meaning. Cambridge, MA: Harvard University Press; 1990

[64] Bruner J. The Culture of Education. Cambridge, MA: Harvard University Press; 1996

[65] Shagrir O. In defense of the semantic view of computation. Synthese. 2020;**197**:4083-4108. DOI: 10.1007/s11229-018-01921-z

[66] Chirimuuta M. Minimal models and canonical neural computations: The distinctness of computational explanation in neuroscience. Synthese. 2014;**191**(2):127-153

[67] Serban M. The scope and limits of a mechanistic view of computational explanation. Synthese. 2015;**192**:3371-3396

Chapter 2

Learning Robotic Ultrasound Skills from Human Demonstrations

Miao Li and Xutian Deng

Abstract

Robotic ultrasound system plays a vital role in assisting or even replacing sonographers in some cases. However, modeling and learning ultrasound skills from professional sonographers are still challenging tasks that hinder the development of ultrasound systems' autonomy. To solve these problems, we propose a learning-based framework to acquire ultrasound scanning skills from human demonstrations[1]. First, ultrasound scanning skills are encapsulated into a high-dimensional multi-modal model, which takes ultrasound images, probe pose, and contact force into account. The model's parameters can be learned from clinical ultrasound data demonstrated by professional sonographers. Second, the target function of autonomous ultrasound examinations is proposed, which can be solved roughly by the sampling-based strategy. The sonographers' ultrasound skills can be represented by approximating the limit of the target function. Finally, the robustness of the proposed framework is validated with the experiments on ground-true data from sonographers.

Keywords: robotic ultrasound, robotic skills learning, learning from demonstrations, compliant manipulation, multi-modal prediction

1. Introduction

Ultrasound imaging technology is widely used in clinical diagnosis due to its noninvasive, low-hazard, real-time imaging, relative safety, and low cost. Nowadays, ultrasound imaging can quickly detect diseases of different anatomical structures, including liver [1], gallbladder [2], bile duct [3], spleen [4], pancreas [5], kidney [6], adrenal gland [7], bladder [8], prostate [9], and thyroid [10]. Besides, during the global pandemic caused by COVID-19, ultrasound is largely used for the diagnosis of infected persons by detecting pleural effusion [11, 12]. However, the performance of ultrasound examination is highly dependent on the ultrasound skills of sonographers, in terms of ultrasound images, probe pose, and contact force (**Figure 1**). In general, the training of an eligible sonographer requires a relatively large amount of time and cases [13, 14]. In addition, the high-intensity repetitive scanning process causes a

---

[1] More details about our original research: https://arxiv.org/abs/2111.09739; https://arxiv.org/abs/2111.01625.

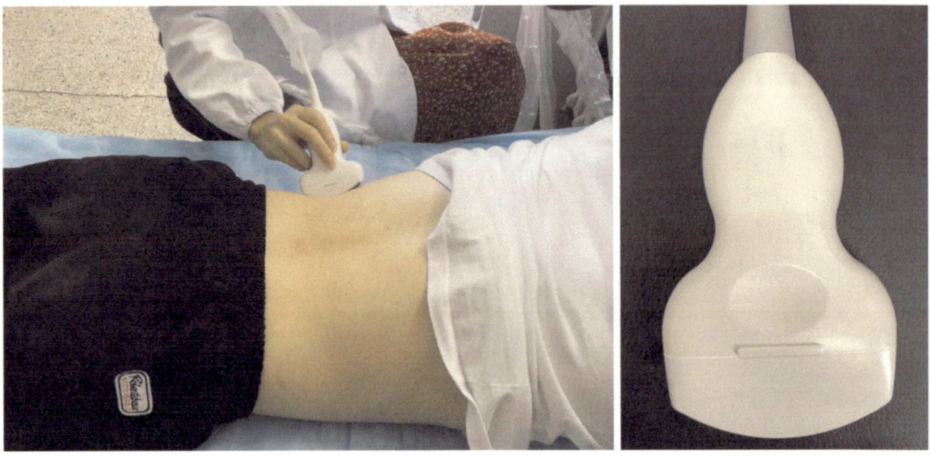

(a) (b)

Figure 1.
The medical ultrasound examination (as left figure shown) needs the dexterous manipulation of ultrasound probe (as right figure shown), which is caused by the environmental complexity in terms of ultrasound images, probe pose and contact force. (a) Clinical medical ultrasound examination. (b) Ultrasound probe.

heavy burden on sonographers' physical condition, further leading to the scarcity of ultrasound practitioners.

To address these issues, many previous studies in robotics have attempted to use robots to help or even replace sonographers [15–17]. According to the extent of the system autonomy, robotic ultrasound can be categorized into three levels—teleoperated, semi-autonomous, and full-autonomous. A teleoperated robotic ultrasound system usually contains two main parts—teacher site and student site [18–20]. The motion of the student robot is completely determined by the teacher, usually a trained sonographer, through different kinds of interaction devices, including a 3D space mouse [18], inertial measurement unit (IMU) handle [20, 21], and haptic interface [21]. While for a semi-autonomous robotic ultrasound system, the motion of the student robot is only partly determined by the teacher [22–24].

For a full-autonomous robotic ultrasound system, the student robot is supposed to perform the whole process of local ultrasound scanning by itself [25–27] and the teacher robot is only used for emergencies. Until today, only part full-autonomous robotic ultrasound system has been reported in the literature [28, 29]. These robotic ultrasound systems usually focus on the scanning of certain anatomical structures, such as the abdomen [28], thyroid [26], and vertebra [29]. A comprehensive survey on robotic ultrasound is given in **Table 1**. Despite these achievements, there are still many obstacles to the development of the robotic ultrasound system. For example, the robustness of most systems is poor and some preparations are required before performing the examination. The key is that there is not a high-dimensional model to learn ultrasound skills (**Figure 2**) from the sonographer, further to guide the adjustment of the ultrasound probe.

In this chapter, we proposed a learning-based approach to represent and learn ultrasound skills from sonographers' demonstrations, and further guide the scanning process [31]. During the learning process, the ultrasound images together with the relevant scanning variables (the probe pose and the contact force) are recorded and encapsulated into a high-dimensional model. Then, we leverage the

Paper	Autonomy degree	Specific target	Modality	Guidance	Publication year
[18]	teleoperated	no	force, orientation, position	human	2015
[19]	teleoperated	no	force, orientation, position	human	2016
[20]	teleoperated	no	force, orientation, position	human	2017
[21]	teleoperated	no	force, orientation, position	human	2020
[22]	semi-autonomous	no	force, orientation, position, elastogram	elastogram, human	2017
[23]	semi-autonomous	no	force, orientation, position, vision	CNN, human	2019
[24]	semi-autonomous	yes	force, orientation, position	trajectory, human	2019
[30]	semi-autonomous	yes	force, orientation, position, image	CNN, human	2020
[25]	full-autonomous	yes	force, orientation, position, vision, image, MRI	vision, MRI, confidence map	2016
[26]	full-autonomous	yes	force, orientation, position, image	SVM	2017
[27]	full-autonomous	no	force, orientation, position, vision	vision	2018
[28]	full-autonomous	yes	force, orientation, position, vision, MRI	vision, MRI	2016
[29]	full-autonomous	yes	force, position, vision	RL	2021

Table 1.
A brief summary of robotic ultrasound. Initials: Convolutional neural network (CNN), magnetic resonance imaging (MRI), support vector machine (SVM), reinforcement learning (RL).

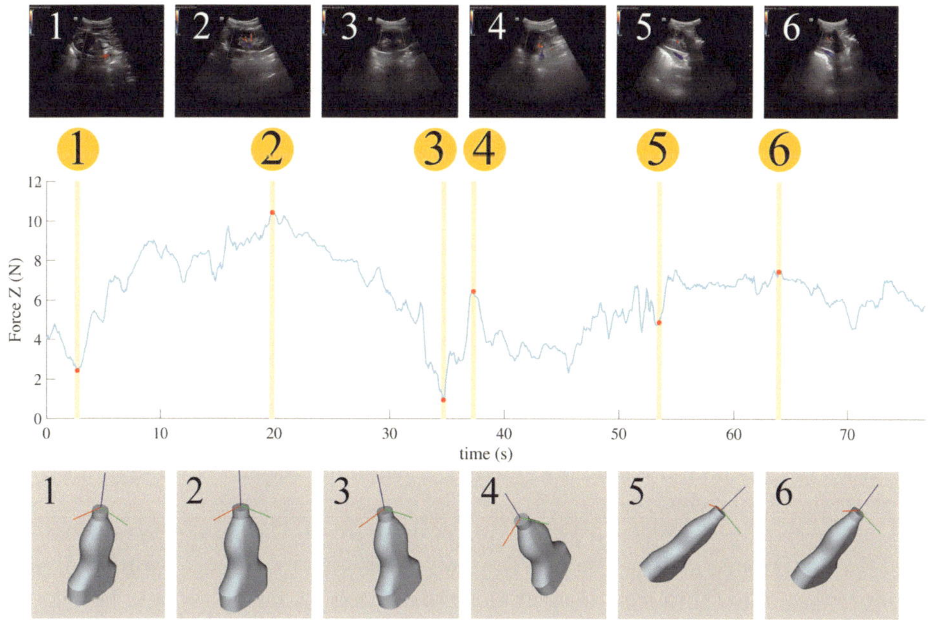

Figure 2.
The feedback information from three different modalities during a free-hand ultrasound scanning process. The first row represents ultrasound images. The second row represents the contact force in the z-axis between the probe and the skin, collected using a six-dimensional force/torque sensor. The third row represents the probe pose, which is collected using an inertial measurement unit (IMU).

power of deep learning to implicitly capture the relation between the quality of ultrasound images and scanning skills. During the execution stage, the learned model is used to evaluate the current quality of the ultrasound image. To obtain a high-quality ultrasound image, a sampling-based approach is used to adjust the probe motion.

The main contribution of this chapter is two-fold: 1. A multi-modal model of ultrasound scanning skills is proposed and learned from human demonstrations, which takes ultrasound images, the probe pose, and the contact force into account. 2. Based on the learned model, a sampling-based strategy is proposed to adjust the ultrasound scanning process, to obtain a high-quality ultrasound image. Note that the goal of this chapter is to offer a learning-based framework to understand and acquire ultrasound skills from human demonstrations [31]. However, it is obvious that the learned model can be ported into a robot system as well, which is our work for the next step [32].

This chapter is organized as follows. Section II presents related work in the field of ultrasound images and ultrasound scanning guidance. Section III provides the methodology of our model, including the learning process of task representation, the data acquisition process through human demonstrations, and the strategy for scanning guidance during real-time execution. Section IV describes the detailed experimental validation, with a final discussion and conclusion in Section V.

2. Related work

2.1 Ultrasound images evaluation

The goal of the ultrasound image evaluation is to understand images in terms of classification [33], segmenting [34], recognition [35], etc. With the rise of deep learning, many studies have attempted to process ultrasound images with the help of neural networks.

Liu et al. have summarized the extensive research results on ultrasound image processing with different network structures, including convolution neural network (CNN), recurrent neural network (RNN), auto-encoder network (AE), restricted Boltzmann's machine (RBM), and deep belief network (DBN) [36]. From the perspective of applications, Sridar et al. have employed CNN for the main plane classification in fetal ultrasound images, considering both local and global features of the ultrasound images [37]. To judge the severity of patients, Roy et al. have collected ultrasound images of the COVID-19 patient's lesions to train a spatial transformer network [38]. Deep learning is also adopted in the task of segmenting thyroid nodules from real-time ultrasound images [39]. While deep learning provides a superior framework to understand ultrasound images, it generally requires a large number of expert-labeled data, which can be difficult and expensive to collect.

Confidence map provides an alternative method in ultrasound image processing [40]. The confidence map is obtained through pixel-wise confidence estimation using a random walk algorithm. Chatelain et al. have devised a control law based on the ultrasound confidence map [41, 42], with the goal to adjust the in-plane rotation and motion of the probe. Confidence map is also employed to automatically determine the proper parameters for ultrasound scanning [25]. Furthermore, the advantages of the confidence maps have been demonstrated by combining with position control and force control to perform automatic position and pressure maintenance [43].

However, a confidence map is proposed with the hand-coded rules, which can not be directly used to guild the scanning motion.

2.2 Learning of the ultrasound scanning skills

While the goal of ultrasound image processing is to understand images, learning ultrasound scanning skills aims to obtain high-quality ultrasound images through the adjustment of the scanning operation. Droste et al. have used a clamping device with IMU to obtain the relation between the probe pose and the ultrasound images during ultrasound examination [44]. Li et al. have built a simulation environment based on 3D ultrasound data acquired by a robot arm mounted with an ultrasound probe [45]. However, they did not explicitly learn ultrasound scanning skills. Instead, a reinforcement learning framework is adopted to optimize the confidence map of ultrasound images, by adapting the movement of the ultrasound probe. All of the above-mentioned work only take the pose and the position of the probe as input, while in this chapter, the contact force between the probe and humans is also encoded, which is considered as a crucial factor during the ultrasound scanning process [46].

For the learning of force-relevant skills, a great variety of previous studies in robotic manipulation focused on learning the relation between force information and other task-related variables, such as the position and velocity [47], the surface electromyography [48], the task states and constraints [49], and the desired impedance [50–52]. A multi-modal representation method for contact-rich tasks has been proposed in ref. [53] to encode the concurrent feedback information from vision and touch. The method was learned through self-supervision, which can be further exploited to improve the sampling efficiency and the task success rate. To the best of our knowledge, for a multi-modal manipulation task, including feedback information from ultrasound, force, and motion, this is the first work to learn the task representation and the corresponding manipulation skills from human demonstrations.

3. Problem statement and method

Our goal is to learn free-hand ultrasound scanning skills from human demonstrations. We want to evaluate the multi-modal task quality of combining multiple sensory information, including ultrasound images, the probe pose, and the contact force, with the goal to extract skills from the task representation and even transferring skills across tasks. We formulate the multisensory data by a neural network, where the parameters are trained by the data supervised by human ultrasound experts. In this section, we will discuss the learning process of the task representation, the data collection procedure, and the online ultrasound scanning guidance respectively.

3.1 Learning of ultrasound task representation

For a free-hand ultrasound scanning task, three types of sensory feedback are available—ultrasound images from the ultrasound machine, force feedback from a mounted F/T sensor, and the probe pose from a mounted IMU. To encapsulate the heterogeneous nature of this sensory data, we propose a domain-specific encoder to model the task, as shown in **Figure 3**. For the ultrasound imaging feedback, we use a VGG-16 network to encode the $224 \times 224 \times 3$ RGB images and yield a 128-d feature vector. For the force and pose feedback, we encode them with a four-layer fully

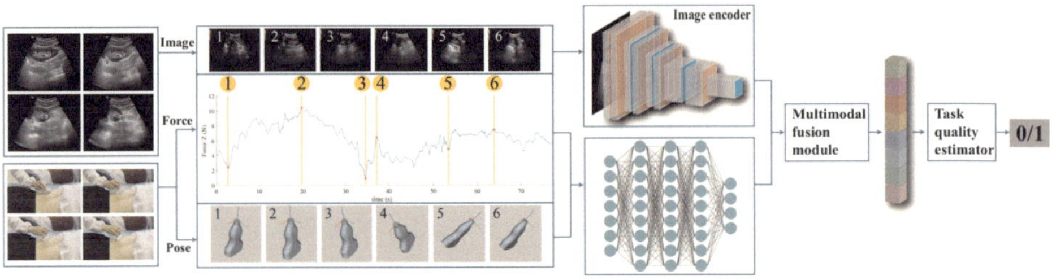

Figure 3.
The multi-modal task learning architecture with human annotations. The network takes data from three different sensors as input—The ultrasound images, force/torque (F/T), and the pose information. The data for the task learning is acquired through human demonstrations, where the ultrasound quality is evaluated by sonographers. With the trained network, the multi-modal task can be represented as a high-dimensional vector.

connected neural network to produce a 128-d feature vector. The resulting two feature vectors are concatenated together into one 256-d vector and connected with a one-layer fully connected network to yield a 128-d feature vector as the *task feature vector*. The multi-modal task representation is a neural network model denoted by Ω_θ, where the parameters are trained as described in the following section.

3.2 Data collection via human demonstration

The multi-modal model as shown in **Figure 3** has a large number of learnable parameters. To obtain the training data, we design a procedure to collect the ultrasound scanning data from human demonstrations, as shown in **Figure 4**. A novel probe holder is designed with intrinsically mounted sensors such as IMU and F/T sensors. A sonographer is performing the ultrasound scanning process with the probe, and the data collected during the scanning process is described as follows:

- $D = \{(S^i, P^i, F^i)\}_{i=1\ldots N}$ denotes a dataset with N observations.
- $S^i \in \mathbb{R}^{224\times224\times3}$ denotes the i-th collected ultrasound image with cropped size.

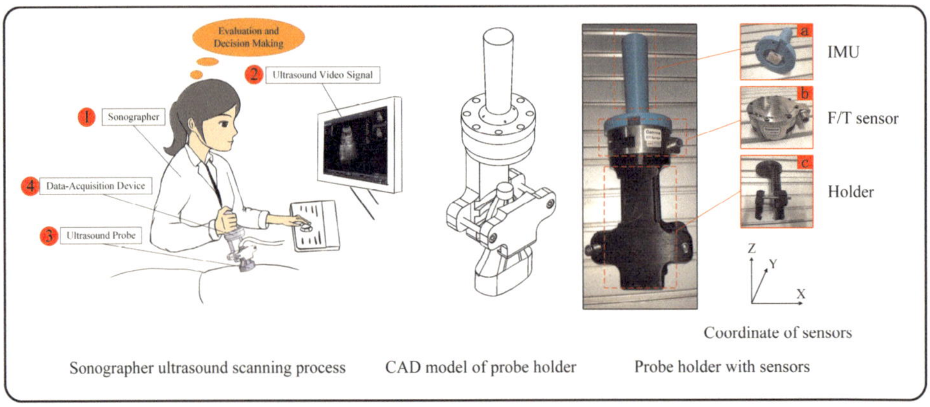

Figure 4.
The ultrasound scanning data collected from human demonstrations. The sonographer is performing an ultrasound scanning with a specifically designed probe holder. The sensory feedback during the scanning process is recorded, including the ultrasound images from an ultrasound machine, the contact force and torque from a 6D F/T sensor, and the probe pose from an IMU sensor.

- $P^i \in \mathbb{R}^4$ denotes the probe pose in terms of quaternion.

- $F^i \in \mathbb{R}^6$ denotes the i-th contact force/torque between the probe and the human skin.

For each recorded data in the dataset D, the quality of the obtained ultrasound image is evaluated by three sonographers and labeled with 1/0. 1 stands for a good ultrasound image while 0 corresponds to an unacceptable ultrasound image. With the recorded data and the human annotations, the model Ω_θ is trained with a loss function of cross-entropy. During training, we minimize the loss function with stochastic gradient descent. Once trained, this network produces a 128-d feature vector and evaluates the quality of the task at the same time. Given the task representation model Ω_θ, an online adaptation strategy is proposed to improve the task quality by leveraging the multi-modal sensory feedback, as discussed in the next section.

3.3 Ultrasound skill learning

As discussed in related work, it is still challenging to model and plan complex force-relevant tasks, mainly due to the inaccurate state estimation and the lack of a dynamics model. In our case, it is difficult to explicitly model the relations among ultrasound images, the probe pose, and the contact force. Therefore, we formulate the policy of ultrasound skills as a model-free reinforcement learning problem, and the target function is as follows:

$$\begin{aligned} \underset{P,F}{\text{maxmize}} \quad & Q_\theta = f(S, P, F|\Omega_\theta) \\ \text{subject to} \quad & P \in D_P, \;\; F \in D_F, \\ & F_z \geq 0. \end{aligned} \tag{1}$$

where Q_θ denotes the quality of the task, which is computed using the learned model Ω_θ by passing through the sensory feedback S, P, F. The constraint $F_z \geq 0$ means that the contact force along the normal direction should be positive. D_P and D_F denote feasible sets of the probe pose and the contact force, respectively. In our case, these two feasible sets are determined by human demonstrations. However, it is worth mentioning that other task-specific constraints for the pose and the contact force can also be adopted here.

By choosing model-free, it requires no prior knowledge of the dynamics model of the ultrasound scanning process, namely the transition probabilities from one state (current ultrasound image) to another (next ultrasound image). More specifically, we choose Monte Carlo policy optimization [54], where the potential actions are sampled and selected directly from previous demonstrated experience, as shown in **Figure 5**. For the sampling, we impose a bound between P'_t, F'_t and P_t, F_t, which prevents the next state from moving too far away from the current state. If the new state $< P'_t, F'_t, S_t >$ is evaluated by the task quality function Q_θ as good, thus the desired pose P'_t and contact force F'_t are used as a goal for the human ultrasound scanning guidance. Otherwise, new P'_t and F'_t are sampled from the previous demonstrated experience. This process repeats $\mathcal{N}$ times, and the P'_t, F'_t with the best task quality, is

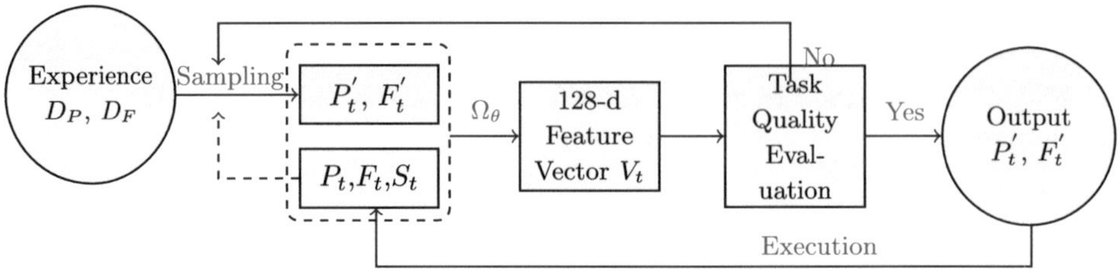

Figure 5.
Our strategy for scanning guidance takes the current pose P_t, the contact force F_t, and the ultrasound image S_t as input, and outputs the next desired pose P_t' and contact force F_t'. For sampling, we impose a bound between P_t', F_t', and P_t, F_t, which prevents the next state from moving too far away from the current state. For evaluation, if the sampled pose and force are predicted as high-quality according to Eq. 1, the skill-learned model will select them as desired output, otherwise, it will repeat the sampling process. For execution, the desired pose P_t' and contact force F_t' are used as the goal for the human ultrasound scanning guidance.

chosen as the final goal for the human scanning guidance. Note that this sampling-based approach does not guarantee the global optimality of Eq. 1. However, this is sufficient for human ultrasound scanning guidance because the final goal is only required to be updated at a relatively low frequency.

4. Experiments: design and results

In this section, we use real experiments to examine the effectiveness of our proposed approach to multi-modal task representation learning. In particular, we design experiments to verify the following two questions:

- Does the force modality contribute to task representation learning?
- Is the sampling-based policy effective for real data?

4.1 Experiments setup

For the experimental setup, we used a Mindray DC-70 ultrasound machine with an imaging frame rate of 900 Hz. The ultrasound image was captured using MAGEWELL USB Capture AIO with a frame rate of 120 Hz and a resolution of 2048 × 2160, as shown in **Figure 6**.

As shown in **Figure 4**, the IMU mounted on the ultrasound probe was ICM20948 and the microcontroller unit (MCU) was STM32F411. The highest frequency of IMU could reach 200 Hz, with an acceleration accuracy of 0.02 g and a gyroscope accuracy of 0.06°/s. The IMU could output the probe pose in the forms of quaternion. For the force feedback, we used a 6D ATI Gamma F/T sensor with a maximum frequency of 7000 Hz. The computer used for the data collection was with Intel i5 CPU and Nvidia GTX 1650 GPU, and with the operating system of Ubuntu16.04 LTS and ROS Kinetic.

4.2 Data acquisition

To make collected data comparable, the recording program needs to implement two functions—coordinate transformation and gravity compensation. The IMU will start to work as soon as the power is turned on. At that time, the probe pose corresponds to the initial coordinate system, so the quaternion's values are equal to (1, 0, 0, 0) and the

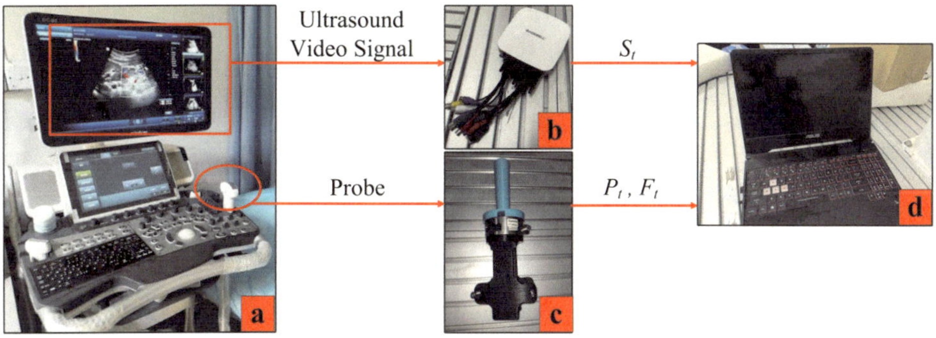

Figure 6.
Experiments setup. (a) the ultrasound machine – Mindray DC-70. (b) the video capture device – MAGEWELL USB capture AIO. (c) Data-acquisition probe holder. (d) the computer for data collection with Intel i5 CPU and Nvidia GTX 1650 GPU, Ubuntu16.04 LTS.

rotation matrix is the identity matrix. However, it will take some time from the wiring of the whole system to recording data, that is, the quaternion's values at the beginning of recording are never equal to the initial ones. To solve this problem, the coordinate transformation is necessary so that the original pose corresponds to the initial coordinate system. Besides, the force/torque signal contains the contact force with the device's gravity, which means our program should have the function of gravity compensation.

The real-time quaternion Q output by the IMU includes four values (w, x, y, z), which should be transformed into a real-time rotation matrix R for calculation. The initial rotation matrix is recorded as R_0. As the rotation matrix is always orthogonal, the inverse and transpose of R_0 are equal and recorded as R_0^{-1}. The relative real-time rotation matrix R_x^* is calculated as follows:

$$R_x^* = R_0^{-1} \cdot R_x \tag{2}$$

The gravity components G_x, G_y, G_z in X, Y, Z directions are calculated by R_x^* and gravity G, as follows:

$$\left[G_x, G_y, G_y\right] = [0, 0, G] \cdot R_x^* \tag{3}$$

In this experiment, we mainly consider the influence of force, so simply record original values of torque. The force/torque sensor's output signal contains real-time force components F_x, F_y, F_z and torque components T_x, T_y, T_z in three directions. The fixed values $F_x^*, F_y^*, F_z^*, T_x^*, T_y^*, T_z^*$ are calculated, as follows:

$$\left[F_x^*, F_y^*, F_z^*\right] = \left[F_x, F_y, F_z\right] - \left[G_x, G_y, G_z\right] \tag{4}$$

$$= \left[T_x, T_y, T_z\right] \tag{5}$$

It is worth noting that gravity G can be calculated by Eq. 6, where the maximum and minimum values of force components in three directions are denoted by $F_{x-max}, F_{x-min}, F_{y-max}, F_{y-min}, F_{z-max}, F_{z-min}$.

$$G = \frac{F_{x-max} - F_{x-min} + F_{y-max} - F_{y-min} + F_{z-max} - F_{z-min}}{6} \tag{6}$$

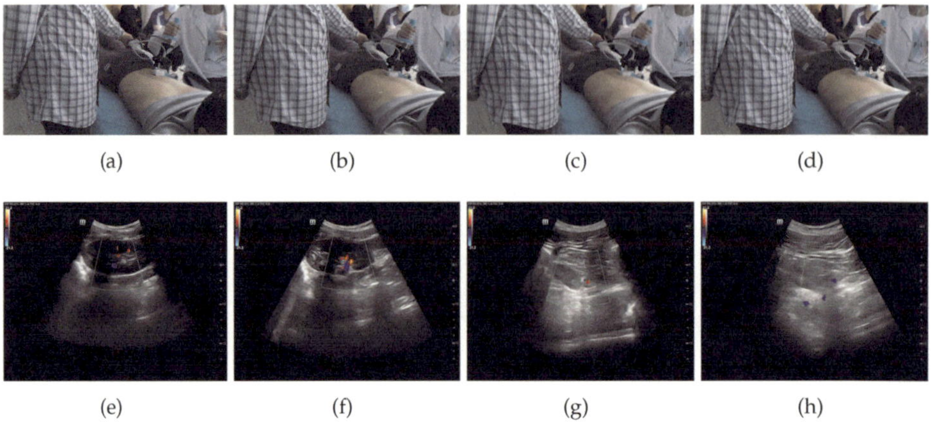

Figure 7.
The snapshots of human ultrasound scanning demonstrations and samples of the obtained ultrasound images. Here the images (e) and (f) are labeled as good quality while (g) and (h) are labeled as bad quality.

The recording frequency is 10 Hz and the accuracy of gravity compensation is 0.5 N. The ultrasound data were collected at the Hospital of Wuhan University. The sonographer was asked to scan the left kidneys of 5 volunteers with different physical conditions. Before the examination, the sonographer vertically held the probe above the left kidney of a volunteer. The ultrasound scanning process began with the recording program launched. The snapshots for the scanning process are shown in **Figure 7**. The collected data consists of ultrasound videos, the probe pose (quaternion), the contact force (force and torque), and labels (1/0). In total, there are 5995 samples of data. The number of positive samples (labeled 1) is 2266, accounting for 37.8%. The number of negative samples (labeled 0) is 3729, accounting for 62.2%. **Figure 8** presents trajectories of the recorded information.

4.3 Experimental results

The detailed architecture of our network is shown in **Figure 9**. In this case, the 256-dimensional vector denotes the feature vector presented in **Figure 3**. We started the training process with a warm start to classify the ultrasound images. The adopted neural network was VGG-16 with cross-entropy loss. A totla of 5995 sets of recorded data were divided into 8:2 for training and validation. Data for training included ultrasound images and labels. The learning rate was 0.001 and the batch size was 20. For the ultrasound skill evaluation, data for training included images S, quaternion P, force F, and labels. By inputting P, F, S, this neural network would output predicted label. We fixed channels of the last fully connected layer in VGG-16 to 128 channels and merged it with (P, F) feature vector. Four fully connected layers were added to transform (P, F) vector into 128 channels, which were concatenated with VGG-16 output vector. After getting the vector with 256 channels, two fully connected layers and a softmax layer were added to output the confidence of the label. **Figure 10** presents accuracy and loss in training neural networks. The neural network for classification finally reached an accuracy of 96.89% and 95.61% in training and validation. The neural network for ultrasound skill

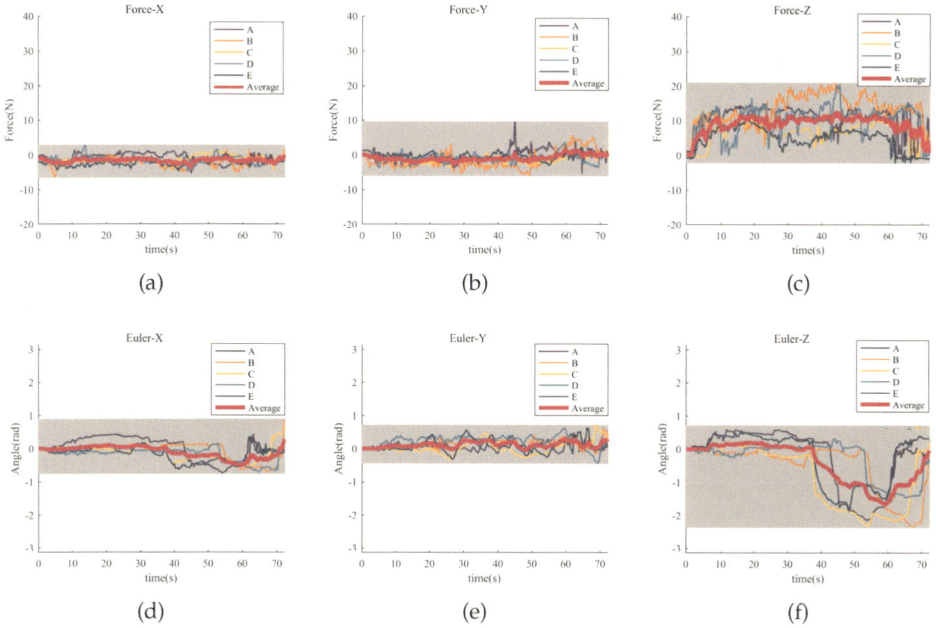

Figure 8.
The trajectories of the recorded force and pose during an ultrasound examination. Force component in (a) X direction (b) Y direction (c) Z direction; rotation axis: (d) X Axis (e) Y Axis (f) Z Axis.

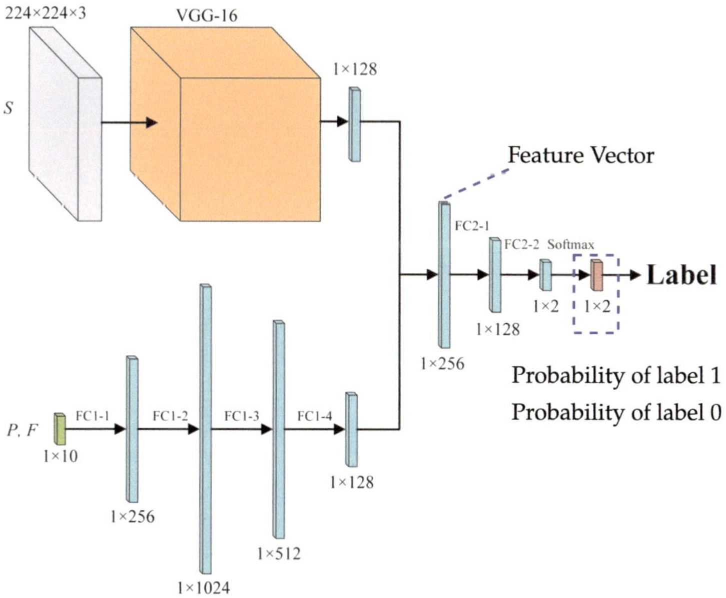

Figure 9.
Framework of the neural network. The ultrasound images were encoded with VGG-16. Four fully connected layers were added to transform (P, F) the vector into 128 channels. Vectors from S and (P, F) were concatenated. Two fully connected layers were added to transform concatenated vector's channels from 256 to 2. Finally, the softmax layer would map the last values to the probability of label 1 or 0.

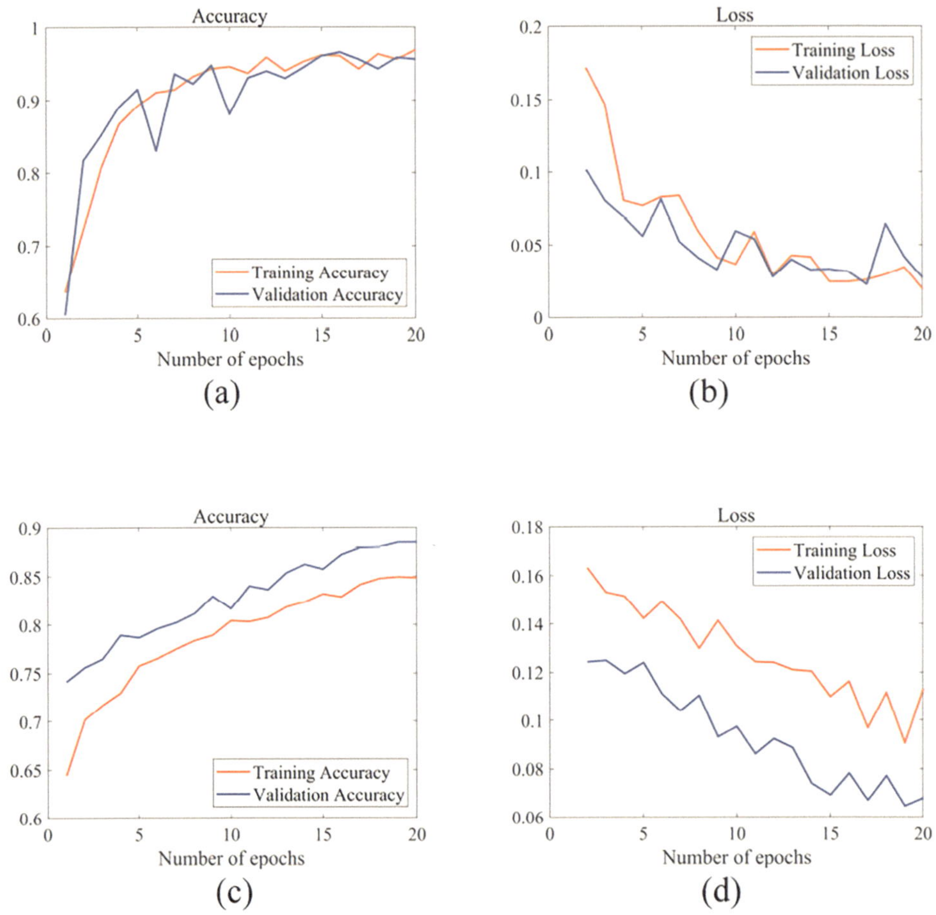

Figure 10.
(a) Accuracy and (b) loss in training the neural network for ultrasound image classification. (c) Accuracy and (d) loss in training the neural network for ultrasound skills evaluation.

evaluation finally reached an accuracy of 84.85% and 88.50% in training and validation.

To confirm the correlation between P and F, we divided data into different levels for training of four networks with different input ports. Net1 was trained with S and P, while Net2 was trained with S and F. Net3 was trained with S, P, and F with two parallel four-layer fully connected neural networks for inputting P and F. Net4 (**Figure 9**) was trained with S, P, and F, with concatenated (P, F) vectors. The main difference between Net3 and Net4 was the existence of interactions between P and F during the training process. Each network had been trained five times with 20 training epochs. **Figure 11** presents the performance of four networks in validation.

Online ultrasound scanning skill guidance: We selected some continuous data streams from the dataset for verification, which had not been used for training the neural network. The sampling process in **Figure 5** was repeated 1000 times and the actions P, F with the best task quality were selected as the next desired action. The whole process took 3 to 5 seconds to output the desired action.

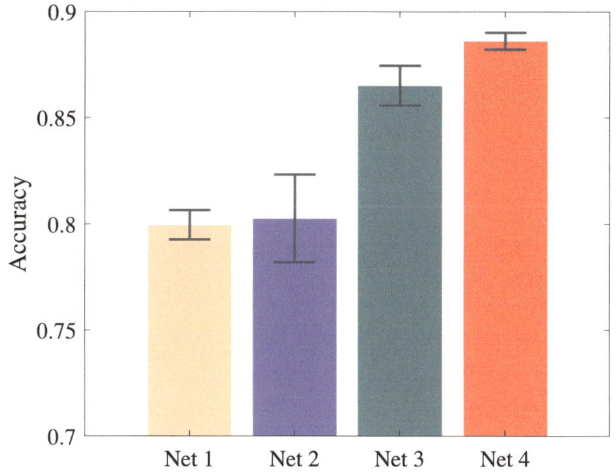

Figure 11.
Accuracy of four networks in validation. Net1 was trained with S and P. Net2 was trained with S and F. Net3 was trained with S, P, and F, without interaction between P and F. Net4 was trained with S, P, and F, with the interaction between P and F.

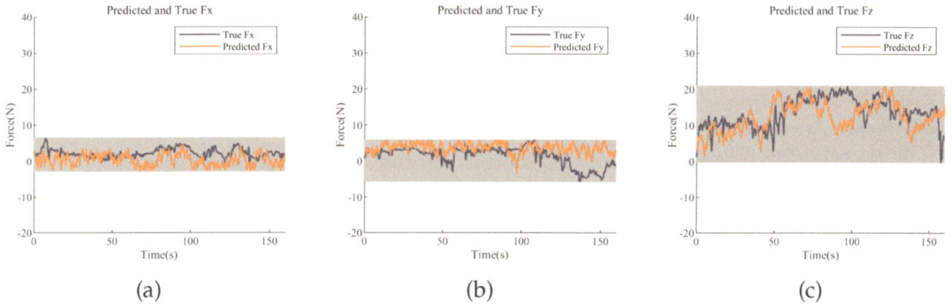

Figure 12.
Predicted force's component in (a) X-axis direction. (b) Y-axis direction. (c) Z-axis direction.

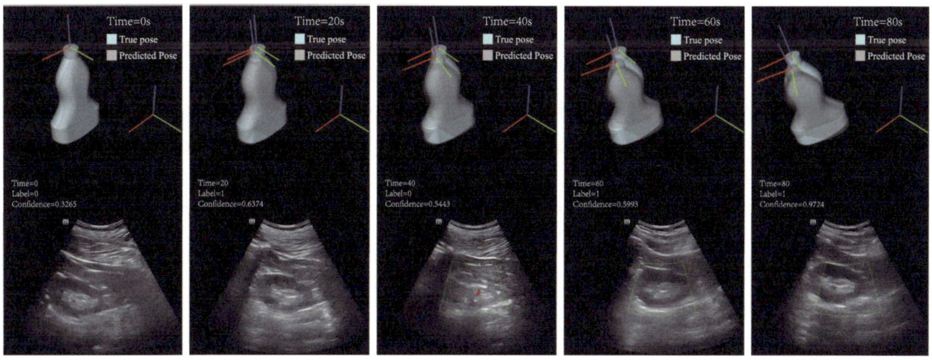

Figure 13.
Predicted probe pose and corresponding ultrasound images. The confidence is the probability of label 1.

Figure 12 presents predicted results about components of contact force, compared with ground truth data. **Figure 13** presents the predicted probe pose with corresponding ultrasound images. **Figure 14** presents predicted and true probe poses with corresponding ultrasound images.

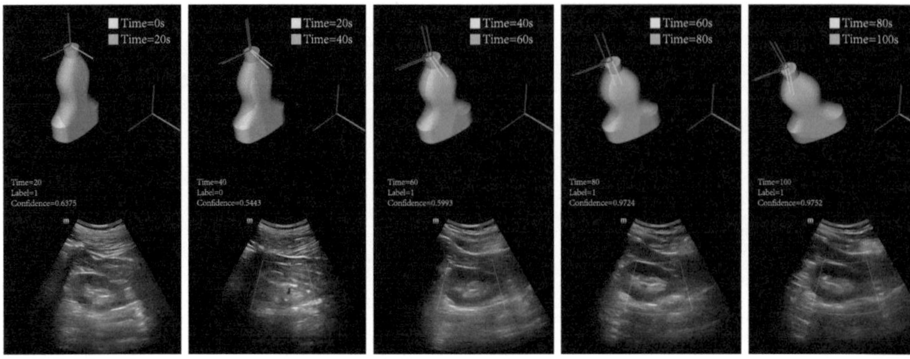

Figure 14.
Predicted and true probe pose, with corresponding ultrasound images. The confidence is the probability of label 1.

5. Discussion and conclusion

5.1 Discussion

This chapter provides a general approach to realizing autonomous ultrasound guidance with some merits as follows: (1) The clinical ultrasound skills are considered as a multi-modal model without any unique factor or parameter, namely, it could be used in most robotic ultrasound systems. (2) The ultrasound skills are mapped into low-dimensional vectors, which makes our approach more flexible with other machine learning methods, such as support vector machine, Gaussian mixture model, and k-nearest neighbors algorithm. (3) The autonomous ultrasound examinations are defined as roughly solving the proposed target function by Monte Carlo method, which provides a newborn and robust method to fulfill autonomous ultrasound.

There are some limitations in this chapter. First, the online guidance method is based on random sampling, which leads to a certain degree of randomness. Therefore, there is a certain difference between forecast results and true values in the short term. Second, to ensure the effectiveness of the sampling, a large number of samples are required, which means a higher task quality improvement would require more computation cost. With the expedition of the dataset, this method is difficult to meet the requirement of timely guidance, which can be solved by denoting the feasible set as a probabilistic model to acquire better sampling efficiency. Finally, we believe that through detailed adjustments to the neural network, the efficiency of this model has the opportunity to be greatly improved without losing too much accuracy.

6. Conclusion

This chapter presents a framework for learning ultrasound scanning skills from human demonstrations. By analyzing the scanning process of sonographers, we define the entire scanning process as a multi-modal model of interactions between ultrasound images, the probe pose, and the contact force. A deep-learning-based method is proposed to learn ultrasound scanning skills, from which the skill-representing target function with a sampling-based strategy for ultrasound examination guidance is proposed. Experimental results show that this framework for ultrasound scanning guidance is robust, and presents the possibility of developing a real-time learning guidance system. In future work, we will speed up the prediction process by taking advantage of self-supervision, with the goal to port the learned guidance model into a real robot system.

Acknowledgements

This work was supported by Suzhou key industrial technology innovation project (SYG202121), and the Natural Science Foundation of Jiangsu Province (Grant No. BK20180235).

Author details

Miao Li[1,2*] and Xutian Deng[1,2]

1 Institute of Technological Sciences, Wuhan University, Wuhan, China

2 School of Power and Mechanical Engineering, Wuhan University, Wuhan, China

*Address all correspondence to: limiao712@gmail.com

References

[1] Gerstenmaier J, Gibson R. Ultrasound in chronic liver disease. Insights Into Imaging. 2014;**5**(4):441-455

[2] Konstantinidis IT, Bajpai S, Kambadakone AR, Tanabe KK, Berger DL, Zheng H, et al. Gallbladder lesions identified on ultrasound. Lessons from the last 10 years. Journal of Gastrointestinal Surgery. 2012;**16**(3): 549-553

[3] Lahham S, Becker BA, Gari A, Bunch S, Alvarado M, Anderson CL, et al. Utility of common bile duct measurement in ed point of care ultrasound: A prospective study. The American Journal of Emergency Medicine. 2018;**36**(6):962-966

[4] Omar A, Freeman S. Contrast-enhanced ultrasound of the spleen. Ultrasound. 2016;**24**(1):41-49

[5] Larson MM. Ultrasound imaging of the hepatobiliary system and pancreas. Veterinary Clinics: Small Animal Practice. 2016;**46**(3):453-480

[6] Correas J-M, Anglicheau D, Joly D, Gennisson J-L, Tanter M, Hélénon O. Ultrasound-based imaging methods of the kidney—Recent developments. Kidney International. 2016;**90**(6): 1199-1210

[7] Dietrich C, Ignee A, Barreiros A, Schreiber-Dietrich D, Sienz M, Bojunga J, et al. Contrast-enhanced ultrasound for imaging of adrenal masses. Ultraschall in der Medizin-European Journal of Ultrasound. 2010; **31**(02):163-168

[8] Daurat A, Choquet O, Bringuier S, Charbit J, Egan M, Capdevila X. Diagnosis of postoperative urinary retention using a simplified ultrasound bladder measurement. Anesthesia & Analgesia. 2015;**120**(5):1033-1038

[9] Mitterberger M, Horninger W, Aigner F, Pinggera GM, Steppan I, Rehder P, et al. Ultrasound of the prostate. Cancer Imaging. 2010;**10**(1):40

[10] Haymart MR, Banerjee M, Reyes-Gastelum D, Caoili E, Norton EC. Thyroid ultrasound and the increase in diagnosis of low-risk thyroid cancer. The Journal of Clinical Endocrinology & Metabolism. 2019;**104**(3):785-792

[11] Buonsenso D, Pata D, Chiaretti A. Covid-19 outbreak: Less stethoscope, more ultrasound. The Lancet Respiratory Medicine. 2020;**8**(5):e27

[12] Soldati G, Smargiassi A, Inchingolo R, Buonsenso D, Perrone T, Briganti DF, et al. Proposal for international standardization of the use of lung ultrasound for covid-19 patients; a simple, quantitative, reproducible method. Journal of Ultrasound in Medicine. 2020;**10**:1413-1419

[13] Arger PH, Schultz SM, Sehgal CM, Cary TW, Aronchick J. Teaching medical students diagnostic sonography. Journal of Ultrasound in Medicine. 2005;**24**(10): 1365-1369

[14] Hertzberg BS, Kliewer MA, Bowie JD, Carroll BA, DeLong DH, Gray L, et al. Physician training requirements in sonography: How many cases are needed for competence? American Journal of Roentgenology. 2000;**174**(5):1221-1227

[15] Boctor EM, Choti MA, Burdette EC, Webster RJ III. Three-dimensional ultrasound-guided robotic needle placement: An experimental evaluation.

The International Journal of Medical Robotics and Computer Assisted Surgery. 2008;**4**(2):180-191

[16] Priester AM, Natarajan S, Culjat MO. Robotic ultrasound systems in medicine. IEEE Transactions on Ultrasonics, Ferroelectrics, and Frequency Control. 2013;**60**(3):507-523

[17] Chatelain P, Krupa A, Navab N. 3d ultrasound-guided robotic steering of a flexible needle via visual servoing. In: 2015 IEEE International Conference on Robotics and Automation (ICRA). Washington, USA: IEEE; 2015. pp. 2250-2255

[18] Seo J, Cho J, Woo H, Lee Y. Development of prototype system for robot-assisted ultrasound diagnosis. In: 2015 15th International Conference on Control, Automation and Systems (ICCAS). Busan, Korea: IEEE; 2015. pp. 1285-1288

[19] Mathiassen K, Fjellin JE, Glette K, Hol PK, Elle OJ. An ultrasound robotic system using the commercial robot ur5. Frontiers in Robotics and AI. 2016;**3**:1

[20] Guan X, Wu H, Hou X, Teng Q, Wei S, Jiang T, et al. Study of a 6dof robot assisted ultrasound scanning system and its simulated control handle. In: 2017 IEEE International Conference on Cybernetics and Intelligent Systems (CIS) and IEEE Conference on Robotics, Automation and Mechatronics (RAM). Ningbo, China: IEEE; 2017. pp. 469-474

[21] Sandoval J, Laribi MA, Zeghloul S, Arsicault M, Guilhem J-M. Cobot with prismatic compliant joint intended for doppler sonography. Robotics. 2020; **9**(1):14

[22] Patlan-Rosales PA, Krupa A. A robotic control framework for 3-d quantitative ultrasound elastography. In: 2017 IEEE International Conference on Robotics and Automation (ICRA). Marina Bay, Singapore: IEEE; 2017. pp. 3805-3810

[23] Mathur B, Topiwala A, Schaffer S, Kam M, Saeidi H, Fleiter T, et al. A semi-autonomous robotic system for remote trauma assessment. In: 2019 IEEE 19th International Conference on Bioinformatics and Bioengineering (BIBE). Athens, Greece: IEEE; 2019. pp. 649-656

[24] Victorova M, Navarro-Alarcon D, Zheng Y-P. 3d ultrasound imaging of scoliosis with force-sensitive robotic scanning. In: 2019 Third IEEE International Conference on Robotic Computing (IRC). Naples, Italy: IEEE; 2019. pp. 262-265

[25] Virga S, Zettinig O, Esposito M, Pfister K, Frisch B, Neff T, et al. Automatic force-compliant robotic ultrasound screening of abdominal aortic aneurysms. In: 2016 IEEE/RSJ International Conference on Intelligent Robots and Systems (IROS). Daejeon, Korea: IEEE; 2016. pp. 508-513

[26] Kim YJ, Seo JH, Kim HR, Kim KG. Development of a control algorithm for the ultrasound scanning robot (nccusr) using ultrasound image and force feedback. The International Journal of Medical Robotics and Computer Assisted Surgery. 2017;**13**(2):e1756

[27] Huang Q, Lan J, Li X. Robotic arm based automatic ultrasound scanning for three-dimensional imaging. IEEE Transactions on Industrial Informatics. 2018;**15**(2):1173-1182

[28] Hennersperger C, Fuerst B, Virga S, Zettinig O, Frisch B, Neff T, et al. Towards mri-based autonomous robotic

us acquisitions: A first feasibility study. IEEE Transactions on Medical Imaging. 2016;**36**(2):538-548

[29] Ning G, Zhang X, Liao H. Autonomic robotic ultrasound imaging system based on reinforcement learning. IEEE Transactions on Bio-medical Engineering. 2021;**68**:2787-2797

[30] Kim R, Schloen J, Campbell N, Horton S, Zderic V, Efimov I, et al. Robot-assisted semi-autonomous ultrasound imaging with tactile sensing and convolutional neural-networks. IEEE Transactions on Medical Robotics and Bionics. 2020;**3**:96-105

[31] Deng X, Lei Z, Wang Y, Li M. Learning ultrasound scanning skills from human demonstrations. 2021. arXiv preprint arXiv:2111.09739. 2021. DOI: 10.48550/arXiv.2111.09739

[32] Deng X, Chen Y, Chen F, Li M. Learning robotic ultrasound scanning skills via human demonstrations and guided explorations. 2021. arXiv preprint arXiv:2111.01625. DOI: 10.48550/arXiv.2111.01625

[33] Hijab A, Rushdi MA, Gomaa MM, Eldeib A. Breast cancer classification in ultrasound images using transfer learning. In: 2019 Fifth International Conference on Advances in Biomedical Engineering (ICABME). Tripoli, Lebanon: IEEE; 2019. pp. 1-4

[34] Ghose S, Oliver A, Mitra J, Mart R, Lladó X, Freixenet J, et al. A supervised learning framework of statistical shape and probability priors for automatic prostate segmentation in ultrasound images. Medical Image Analysis. 2013; **17**(6):587-600

[35] Wang L, Yang S, Yang S, Zhao C, Tian G, Gao Y, et al. Automatic thyroid nodule recognition and diagnosis in ultrasound imaging with the yolov2 neural network. World Journal of Surgical Oncology. 2019;**17**(1):1-9

[36] Liu S, Wang Y, Yang X, Lei B, Liu L, Li SX, et al. Deep learning in medical ultrasound analysis: A review. Engineering. 2019;**5**(2):261-275

[37] Sridar P, Kumar A, Quinton A, Nanan R, Kim J, Krishnakumar R. Decision fusion-based fetal ultrasound image plane classification using convolutional neural networks. Ultrasound in Medicine & Biology. 2019; **45**(5):1259-1273

[38] Roy S, Menapace W, Oei S, Luijten B, Fini E, Saltori C, et al. Deep learning for classification and localization of covid-19 markers in point-of-care lung ultrasound. IEEE Transactions on Medical Imaging. 2020; **39**(8):2676-2687

[39] Ouahabi A, Taleb-Ahmed A. Deep learning for real-time semantic segmentation: Application in ultrasound imaging. Pattern Recognition Letters. 2021;**144**:27-34

[40] Karamalis A, Wein W, Klein T, Navab N. Ultrasound confidence maps using random walks. Medical Image Analysis. 2012;**16**(6):1101-1112

[41] Chatelain P, Krupa A, Navab N. Optimization of ultrasound image quality via visual servoing. In: 2015 IEEE International Conference on Robotics and Automation (ICRA). Washington, USA: IEEE; 2015. pp. 5997-6002

[42] Chatelain P, Krupa A, Navab N. Confidence-driven control of an ultrasound probe: Target-specific acoustic window optimization. In: 2016 IEEE International Conference on Robotics and Automation (ICRA).

Stockholm, Sweden: IEEE; 2016. pp. 3441-3446

[43] Chatelain P, Krupa A, Navab N. Confidence-driven control of an ultrasound probe. IEEE Transactions on Robotics. 2017;**33**(6):1410-1424

[44] Droste R, Drukker L, Papageorghiou AT, Noble JA. Automatic probe movement guidance for freehand obstetric ultrasound. In: International Conference on Medical Image Computing and Computer-Assisted Intervention. Lima, Peru: Springer; 2020. pp. 583-592

[45] Li K, Wang J, Xu Y, Qin H, Liu D, Liu L, et al. Autonomous navigation of an ultrasound probe towards standard scan planes with deep reinforcement learning. Xi'an, China: IEEE; 2021:8302–8308. arXiv preprint arXiv:2103.00718

[46] Jiang Z, Grimm M, Zhou M, Hu Y, Esteban J, Navab N. Automatic force-based probe positioning for precise robotic ultrasound acquisition. IEEE Transactions on Industrial Electronics. 2020;**68**:11200-11211

[47] Gao X, Ling J, Xiao X, Li M. Learning force-relevant skills from human demonstration. Complexity. 2019;**2019**: 5262859

[48] Zeng C, Yang C, Cheng H, Li Y, Dai S-L. Simultaneously encoding movement and semg-based stiffness for robotic skill learning. IEEE Transactions on Industrial Informatics. 2020;**17**(2): 1244-1252

[49] Holladay R, Lozano-Pérez T, Rodriguez A. Planning for multi-stage forceful manipulation. Xi'an, China: IEEE; 2021:6556–6562. arXiv preprint arXiv:2101.02679

[50] Li M, Tahara K, Billard A. Learning task manifolds for constrained object manipulation. Autonomous Robots. 2018;**42**(1):159-174

[51] Li M, Yin H, Tahara K, Billard A. Learning object-level impedance control for robust grasping and dexterous manipulation. In: 2014 IEEE International Conference on Robotics and Automation (ICRA). Hong Kong, China: IEEE; 2014. pp. 6784-6791

[52] Li M, Bekiroglu Y, Kragic D, Billard A. Learning of grasp adaptation through experience and tactile sensing. In: 2014 IEEE/RSJ International Conference on Intelligent Robots and Systems. Chicago, USA: IEEE; 2014. pp. 3339-3346

[53] Lee MA, Zhu Y, Srinivasan K, Shah P, Savarese S, Fei-Fei L, et al. Making sense of vision and touch: Self-supervised learning of multimodal representations for contact-rich tasks. In: 2019 International Conference on Robotics and Automation (ICRA). Montreal, Canada: IEEE; 2019. pp. 8943-8950

[54] Sutton RS, Barto AG. Reinforcement Learning: An Introduction. MIT Press; 2018

Chapter 3

Skill Acquisition for Resource-Constrained Mobile Robots through Continuous Exploration

Markus D. Kobelrausch and Axel Jantsch

Abstract

We present a cognitive mobile robot that acquires knowledge, and autonomously learns higher-level abstract capabilities based on play instincts, inspired by human behavior. To this end, we (i) model skills, (ii) model the robot's sensor and actuator space based on elementary physical properties, and (iii) propose algorithms inspired by humans' play instincts that allow the robot to autonomously learn the skills based on its sensor and actuator capabilities. We model general knowledge in the form of competencies (skills) of the mobile robot based on kinematic properties using physical quantities. Thus, by design, our approach has the potential to cover very generic application domains. To connect desired skills to the primitive capabilities of the robot's sensors and actuators, it playfully explores the effects of its actions on its sensory input, thus autonomously learning relations and dependencies and eventually the desired skill. KnowRob is used for knowledge representation and reasoning, and the robot's operation is based on ROS. In the experiments, we use a millirobot, sized 2 cm^2, equipped with two wheels, motion, and distance sensors. We show that our cognitive mobile robot can successfully and autonomously learn elementary motion skills based on a playful exploration of its wheels and sensors.

Keywords: artificial intelligence, autonomous learning systems, cognitive architecture, reinforcement learning, knowledge representation and reasoning, resource-constrained systems, low-energy mobile robots

1. Introduction

Our starting point is a robot with (a) a set of sensors and actuators, (b) tight resource limitations, (c) access to a database that captures general motion-related competencies (e.g. moving along a rectangle or navigating to a target location), and (d) built-in assumptions about physical laws and geometric relations. Our objective is to develop methods that allow the robot to autonomously learn competencies stored in the database.

Initially, the robot does not know the meaning and effect of its sensors and actuators (e.g. if an actuator controls a LED or a wheel). Therefore, the first activities are concerned with learning the meaning of its sensors and the effects of its actuators. Then, basic competencies from the knowledge base are acquired followed by increasingly complex competencies. A priori, the robot only has built-in knowledge of how to interface its sensors and actuators and basic assumptions about physical laws and geometric relations, but not what the sensors and actuators mean or how a specific motion can be accomplished.

Our long term goal is to provide the robot with general methods that allow the robot to work with any kind of sensors and actuators, in any kind of physical environment, and learning any kind of competence, provided it is possible at all (e.g. if the robot has only LEDs but no motors, it cannot learn to move).

We consider this a worthwhile vision because this approach to minimize prior knowledge and assumptions will facilitate very flexible systems that can work with any kind of sensors and actuators, in wheel-equipped or flying robots, on level plains, rocky or grassy surfaces, or even in wet environments. It will allow the use of accurate or inaccurate sensors and actuators, and to adapt to aging and wear-out effects. This approach is general because the only assumptions we make are the laws of kinematics and geometry, the availability of and access to sensors and actuators, the availability of learning methods (e.g., RL), and the availability of a database describing the skills to be learned.

While this is our vision, in this article, we make the further assumption that the robot knows the meaning of its sensors and operates in a two-dimensional plane. Inspired by the play instinct observed in humans and animals we propose exploratory, hierarchical learning. Simple and elementary tasks are tried out and learned first, followed by complex and composite tasks. This means the robot starts by asking if it can move at all, then it tries to learn elementary linear and angular motions, based upon which it studies moving along rectangles and similarly simple shapes. For each learning task, we use Reinforcement Learning (RL) as it matches well the exploratory nature of the robot's setting. The learning tasks are identified based on entries from a knowledge database that describes the motion skills and the hierarchical relation between skills. Specifically, we use the KnowRob knowledge processing system [1], which is designed to provide autonomous robots with the knowledge base for performing motion and manipulation tasks.

In this paper, we propose and demonstrate the Skill Acquisition Method (SAM) for the case of a wheel-equipped tiny robot operating on a smooth, level plain; in future work, we will show that the same techniques generalize to other settings and environments. We evaluate our approach in a simulation environment for a two-wheeled and a four-wheeled mobile robot moving in a two-dimensional space. Experiments show that the system can learn and interpret its basic motion commands and derive complex motions, and finally, it succeeds in driving a rectangle (set of basic motion commands). Our contributions are summarized as follows:

i) We identify a minimal set of prior knowledge mandatory for learning basic movements.

ii) We propose a cognitive system behavior, the Playful Continuous Competence Acquisition (PCCA), that enables the learning and development of skills based on

a) the model of generic competencies (skills),

b) and the system's Sensor and Actuator Space (SAS) grounded in elementary physical properties.

2. Related work

The use of knowledge representation and reasoning in robots has a long tradition, where the *Shakey* robot had already 1984 an internal representation of its environment [2]. Extensive research has been done in robotics and artificial intelligence in recent decades, to which this article mainly refers. Since robots have specific demands on knowledge bases and appropriate methods, e.g., linking abstract knowledge representation and specific control systems, this can be best solved with frameworks explicitly designed for this purpose.

In this context, KnowRob was specifically developed to equip autonomous robots with knowledge and methods (Knowledge Representation and Reasoning (KR&R)) to perform everyday manipulation tasks and to provide an infrastructure for cognitively enabled robots [1, 3, 4]. It represents one of the most advanced knowledge processing systems for robots, which has evolved even further with OPEN-EASE [5], which integrates KNOWROB2 [6], and aims to provide a remote knowledge and reasoning service that offers unprecedented access to the knowledge of autonomous robotic agents performing human-scale manipulation tasks. This seems promising for agents performing such rich human-scale manipulations but also places significant demands on the system's resources, which is crucial for systems with limited resources. Therefore, we use KnowRob as the basis for knowledge processing and representation to take full advantage, but we target the approaches and methods that allow it to be deployed in such tiny systems.

A recent work dealing with the generalization of experience into abstract knowledge for novel situations, entitled Socio-Physical Model of Activities (SOMA [7]), consists of a comprehensive model for connecting physical and social entities that enable flexible execution by robotic agents. Since this representation seems essential, we use a similar approach, keeping our model flat in the first line due to resource-constraints. This limits the flexibility of the application (smaller knowledge base) but is crucial, and we aim for a reasonable trade-off. In this context, we also discuss a set of a small amount of prior knowledge.

RoboEarth has similar goals and approaches to our work [8]. Capabilities are also modeled, where we differ mainly in how they are used. We assume a set of general prior knowledge and basic methods to acquire skills, while their work accepts more complex algorithms to derive specific knowledge. Additionally, we further evaluate and improve skills to achieve continuous development.

Other works also deal with systems that learn semantically from different experiences, taking different approaches [9]. While learning relies on recorded experiences in semantic structures containing high level representations. A key difference in our approach is that we generate skill-specific episodic knowledge through real-time learning methods, leading to knowledge abstraction at an earlier stage. To further leverage this, we define a set of prior knowledge that must be present to enable use in resource-constrained systems.

3. Skill acquisition

In the following, we present SAM, which starts from a set of general assumptions (knowledge and methods) to autonomously acquire and develop specific complex capabilities and aims for generic deployment in resource-constrained systems.

3.1 Overview

SAM (**Figure 1**) consists of various elements structured in layers. The bottom layer reflects the physical part, i.e., the robot and its environment. The layer above hosts the central computational agent, which abstracts the interface to the environment via the SAS. This general and generic interface is deliberately based on the fundamental physical properties of sensors and actuators. Thus, by definition, any environment can be integrated elegantly and efficiently as long as it follows the matching properties, defined in 3.4. Further, the agent has access to the knowledge base and reasoners. SAM follows a cognitive-behavioral architecture to autonomously learn skills using a KR&R methods combined with real-time learning from the physical environment.

3.1.1 Cognitive model

Cognitive models go beyond traditional behavioral models regarding what an entity (robot) knows, how that knowledge is acquired, and how it can be used. As a result, they are becoming increasingly popular in artificial intelligence. They are well suited for implementing highly autonomous systems that exhibit some intelligence and are expected to develop over time. There are several approaches to these models in the literature, particularly in robotics, which attempt to mimic the behavior of intelligent agents based on human cognition. Recent work on a generic form of this, such as the Socio-physical Model of Activities (SOMA) consists of a comprehensive model that combines physical and social entities and allows flexibility of execution by robotic agents through symbolic reasoning [7].

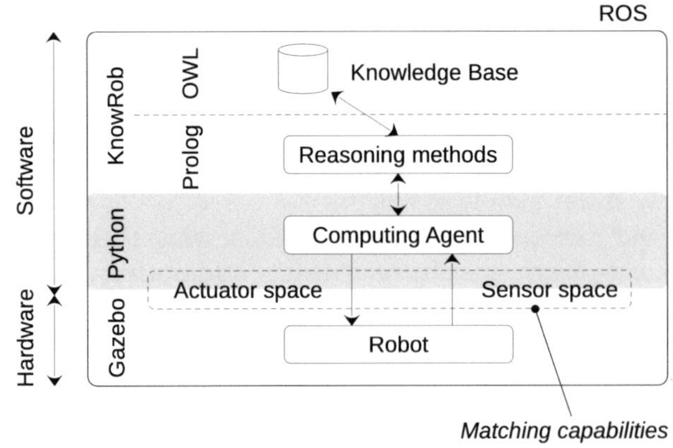

Figure 1.
SAM overview with its layered architecture and the distinction between software and hardware. The KnowRob layer consists of methods for KR&R, while the computing agent (python) drives the system flow to autonomously acquire skills. It has access through the SAS to the physical environment and the database. All components are integrated in ROS.

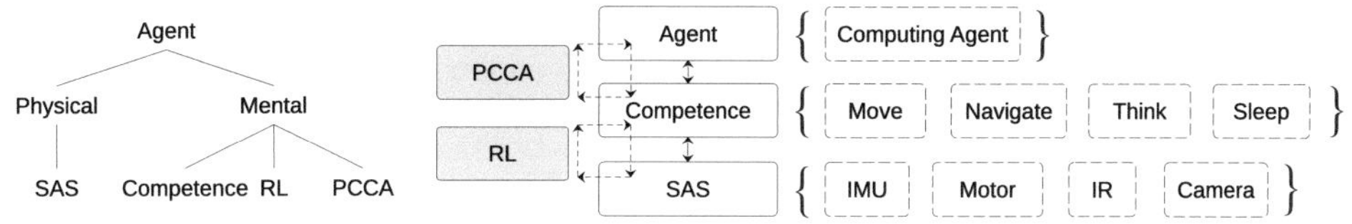

Figure 2.
SAMs cognitive system model. The left side shows the distinction between physical and mental entities. The right side shows their relations, where an agent seeks to acquire and further improve a particular competence using the PCCA method. RL methods are used to learn a specific skill while interacting when required with SAS to access the physical environment. The modules in the dashed line show examples of instances.

Since cognitive models, in the broader sense, represent complex processes and behavior, we focus our modeling on the core elements that we consider essential for our millirobot to acquire basic skills. In the medium term, we intend to adapt them to SOMA.

Figure 2 illustrates our proposed cognitive system model, divided into the physical and mental domains. The SAS depends on the physical properties of the robot. We divide the mental part into elementary capabilities and behavioral methods. A competence reflects knowledge of a particular skill acquired and applied through the PCCA behavioral method, while RL is used to learn a specific competence (e.g. motion commands). We will introduce and discuss these essential elements step by step in this article.

3.1.2 Use of KR&R and episodic memory

For each activity that SAM performs and observes (physical interaction, knowledge inference, learning, etc.), it generates skill-specific knowledge as episode memory and stores it with timestamps. Such episodic memory could include what the robot saw, reasoned, and did, how it did that, why, and what effects it caused [5]. It can be used for further conclusions and learning at any time. While the size and scope of the episodic memory directly relate to the resources required for the particular system. Many approaches attempt to collect a large amount of extensive detailed knowledge, which directly impacts computing time. This seems impractical for systems with limited resources. Therefore, we propose to keep episodic knowledge flat and small and to store only highly relevant information. In this context, we also consider a set of general prior knowledge that an intelligent system must have to learn and exhibit sufficient episodic memory for a given skill. We argue that these two facts are essential to consider for use in systems with tight resource limitations. Section 3.5.1 outlines an approach to a set of concrete prior knowledge and episodic knowledge developed by SAM, intended for use in resource-constrained systems.

Our long-term vision is that all relevant parts of the proposed SAM are hosted on such a system, e.g., a tiny millirobot powered by a micro-controller. We are aware of the challenges of migrating databases and logical reasoning to resource-constrained systems. As an intermediate step, we propose separating the acquisition and exploitation phases, where the system has access to KR&R in the first phase. Once the skill has been successfully acquired (sufficient episodic memory) to some degree, the system may be able to master it independently. Then it exploits the acquired skill with appropriate methods on the tiny millirobot. Whenever the system detects significant changes or decides to search for new capabilities, it contacts the database again. In this way, we can elaborate a similar knowledge acquisition behavior for resource-constrained systems compared to those with fewer constraints that host KR&R directly.

3.2 Competence

A central core element of SAM is competence, generally understood as mental property. The focus of this work is on the modeling of competencies that, when defined, lead to physical actions of the system through the SAS. However, competence in itself does not always have to be related to the physical facts of the system. It could also be a purely mental ability, such as spatial awareness, concentration, attention, reasoning, logic, and so forth. To model capabilities in an intelligent system, essential basic elements of those capabilities must be considered to grant an appropriate developmental progression. In a nutshell, a system should learn a skill independently and reason with appropriate knowledge about how good that skill is. Moreover, the evaluation of skills is of particular interest, used to continuously improve the respective skills. In this way, a cognitive system that also has an interest in developing itself further can become better over time.

In this context, two fundamental elements of competence have been attributed. These are (i) *fitness*, which is a statement of how well system masters the skill, and (ii) *learnability*, which indicates a skill that can be learned by the agent.

Figure 3 illustrates the general concept of competence modeled in the knowledge base. The *fitness* is represented with a numeric value and the *learnability* with a boolean value. The *learnability* is fulfilled if (a) all properties for learning the skill are satisfied, and (b) the system provides methods to learn this competence. The properties of (a) can be determined either by inference knowledge from the database or, if they depend on the physical space, directly by physical interaction. For instance, in the case of the movement skill, we determine the physical agent's ability to move through physical interactions (Section 3.5.1). For (b), certain methods must be in place to learn specific skill knowledge. Such knowledge could be, for example, a set of specific actions and their command values. We use RL to learn specific motion commands executed via SAS. Other learning methods such as Deep RL or supervised/unsupervised methods could also be utilized. However, the goal is to acquire a subset of episodic memory sufficient to exhibit a particular skill. The *fitness* is used to evaluate how well the skill is mastered and is represented by a number from 0 to 100, with 100 being the maximum achievable. For example, we directly assign the RL method reward to *fitness* of a basic motion competence (Section 3.5.4). In addition to the general properties of competence, a corresponding instance may also store specific

Figure 3.
The competence entity with its two fundamental properties (learnability and fitness), modeled in the knowledge base using ontologies.

knowledge relevant to the execution of skill in a particular system. In our case, we memorize the action commands, their fitness, and the timestamp, as discussed in Section 3.5.5.

3.3 Playful continuous competence acquisition

Another key core element is behavior, which ensures continuous development by learning new skills and further improving existing ones. Generally, a system that acquires specific skills should not consider them finally learned after the first success. Instead, the goal is to evaluate what has been learned and, if necessary, to develop further and improve it. In this way, a system can evolve autonomously and continuously adapt to certain changes in its environment. To this end, we consider the following key behavioral elements crucial: (a) the striving for new skills and (b) the continuous improvement of already learned skills.

Figure 4 illustrates our proposed PCCA method, focusing on knowledge acquisition and skill development. An interpretation of the learned skills in terms of possible application scenarios and their combinations in specific contexts, i.e., for which purpose skill could be used, is future work and not considered here. Further, to generalize the high-level system flow, a promising approach would be to model it directly in the knowledge base in tasks and actions. For that, KnowRob offers a promising approach that might also be applicable to our system [1, 7].

However, SAM's high-level behavioral process is determined using the PCCA reasoner, directly queried by the computing agent. We define two different high-level-behavioral phases acting on the competence model properties (*fitness* and *learnability*), shown in **Figure 4**: *(i) seeking for new competencies* and *(ii) improving known competencies*. Phase *(i)* and *(ii)* are general cognitive-behavioral patterns based on the competence model (presented in Section 3.2) that are independent of the skill being learned. Whereas skill-specific learning methods (dashed lines in **Figure 4**),

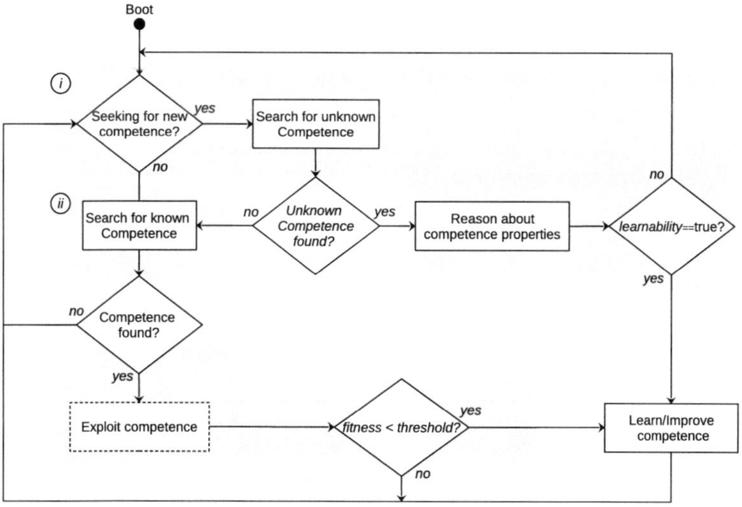

Figure 4.
The PCCA flow is divided into two high-level-behavioral phases, (i) and (ii). It acts based on the competence properties (fitness and learnability). It covers key behavioral elements (a) and (b) by utilizing phases (i) seeking for new competencies and (ii) improving known competencies, in an incremental fashion. The elements marked by dashed lines represent skill-specific learning methods.

triggered by the PCCA, acquire the respective competence-specific knowledge (e.g. an RL element for a navigation skill). It switches playfully between these two phases and can thus develop and improve over time.

3.3.1 Seeking new competencies

SAM searches for new capabilities based on the instances available in the knowledge base. Currently, these still need to be instantiated manually, with the long-term goal being to create them automatically. If one is present, the system uses the competence's *fitness* property to determine if it is already known and learned. If not, the *learnability* property is used to determine if it can be learned. If yes, it enters the skill-specific learning phase, and otherwise, it continues searching.

3.3.2 Improving known competencies

The system decides whether a competence can still be improved based on the *fitness* property. When the *fitness* value is below a certain threshold, SAM relearns the skill by re-running the RL method exploration phase. If a better solution is found, it memorizes it as the best for further use. Moreover, it operates on an incremental basis, ensuring that the best solution is found after a certain period of time. It further allows to react to changes in the environment and thus make immediate adaptations.

3.3.3 Skill-specific learning methods

A specific competence is explored, learned, and exploited using appropriate learning methods (RL, supervised/unsupervised learning). These methods are competence specific and must be designed according to the particular skill. In principle, it is possible to integrate highly optimized learning algorithms for the respective functions. However, our goal is to use basic algorithms and execute them using general knowledge modeled in the knowledge base. In this way, we expect even more flexible usage, where only the primary parameters in the database need to be adjusted while the algorithm remains the same. When needed, the skill-specific learning method is triggered by the PCCA. In Section 3.5.4, we further discuss this approach and propose an RL basic algorithm that we extend with methods from KR&R to achieve generalization.

3.4 Sensor and actuator space

The sensor and actuator space (SAS) represents a generic interface to the robot environment, solely based on physical quantities. For example, consider an Inertial Measurement Unit (IMU), an odometry sensing unit as sensors, and two motors as actuators. SAM's *matchingcapabilities* rely on the physical quantities of those sensors and actuators that the robotic-system must provide. **Figure 5** illustrates the resulting abstracted interfaces for sensors (ψ, x and y) and actuators (m_1 and m_2), with their physical quantities shown in **Table 1**.

We assume that these interfaces abstract the robot-specific sensor data and actuator commands. For example, how the respective motor of the robot is controlled (using a motor controller that takes the acceleration properties into account) needs to

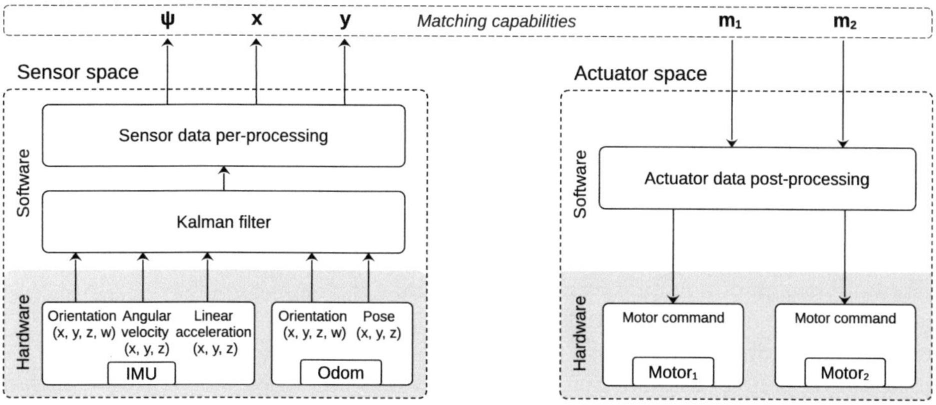

Figure 5.
Example of a sensor and actuator space (SAS) consisting of two motors, an IMU, and an odometry sensing unit. The respective data's pre- and post-processing is robotic-system-dependent and must be addressed individually. Thus, SAS must abstract the low-level data appropriately to meet SAM's matching capabilities.

Data	Description	Physical Quantity	Unit
Sensor space			
ψ	Yaw rotation	Angle	°
x	Horizontal cartesian coordinate	Length	m
y	Vertical cartesian coordinate	Length	m
Actuator space			
m_1	Wheel torque	Torque	N m
m_2	Wheel torque	Torque	N m

Table 1.
Matching capabilities: SAS with its physical quantities.

be modeled robotic-system-dependent. In our case, the respective sensor data pre-processing layer and the actuator data post-processing layer take care of this.

This approach is generic, and we argue that the system initially does not need to know which actuators or sensors it is dealing with. A long-term goal is to employ appropriate methods and knowledge to identify and learn its capabilities. The Semantic Sensor Web follows this approach, annotating sensor data with various semantic metadata (including physical quantities) [10]. Further, there is promising work in automatic semantic knowledge acquisition for sensor data, which aims to annotate raw data with semantic knowledge [11]. Thus, our approach aims to leverage generic interfaces to integrate those methods seamlessly in future work.

However, the specific experimental setup is illustrated in **Figure 5** for a two-wheeled mobile robot. It is equipped with two motors (for a 4-wheeled robot, extended by two additional motors), each driving a wheel, an inertial measurement unit (IMU), and an odometry sensing unit (obtained from the simulation environment) that is used to reduce the drift error of the IMU over time using a Kalman filter [12]. We are well aware of the challenges to the precision of these sensor measurements required for stable localization, which is extensively discussed in many publications [13, 14]. However, we do not further discuss this and assume that the problem is well understood. In conclusion, with this generic design, any robot environment can interact with SAM as long as the required physical *matchingcapabilities* are supported.

3.5 Motion skills

As mentioned earlier, this work focuses on modeling competencies that lead to physical actions of the system through SAS. Considering this fact and the physical characteristics of a wheeled robot, specifically the actuators in the form of wheels, potential movement possibilities can be assumed. For that, we consider basic movements, which in turn are subdivided into atomic and more complex movements. In a broader sense, for atomic actions, the robot is assumed to always be stationary, moving by applying torque to the actuators and stopping when it is removed. Such an atomic motion thus represents a sub-element of a more complex motion. It is not claimed that those movements are the most efficient in terms of smoothness and speed. However, they still allow the robot to approach all positions in a given space. **Figure 6** illustrates a set of motion skills where atomic movements such as angular and linear movements ground complex movement patterns such as rectangles, cycles, or even more generally, a navigation path. The acquisition of these skills occurs in the same hierarchical manner that enhances the physical learning methods discussed in the next section.

3.5.1 Hierarchical knowledge acquisition

Let us first consider the knowledge we can gain about a movement, which we draw from a small set of prior knowledge. Assuming the system has not yet acquired any specific knowledge about motion, it has first to find out whether it can move at all with its given actuators: *(I)"Am I able to move?"* To answer this question, the system initiates random actions and observes their consequences. In our case of a two-wheeled robot, both actuators are moved randomly, and the physical effects are evaluated based on a spatial position change. At the level *(I)* the question is only about the possibility of any movement, as depicted in **Figure** 7. If the system has an actuator that controls only a LED, it would be recognized as irrelevant for movements. Next, at level *(II)*, we can start asking for basic movement patterns without specific lengths or angles. *(II) "Am I able to turn forward/backward/left/right?"* The actuators are triggered again, and SAM searches for angular (left/right) and linear (forward/backward)

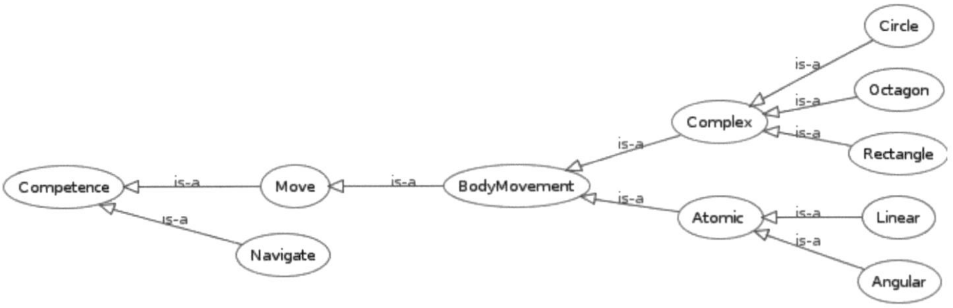

Figure 6.
Competence graph (modeled in the database), with a set of motion skills, sub-divided into atomic and complex. Where the atomic movements form the basis for more complex patterns.

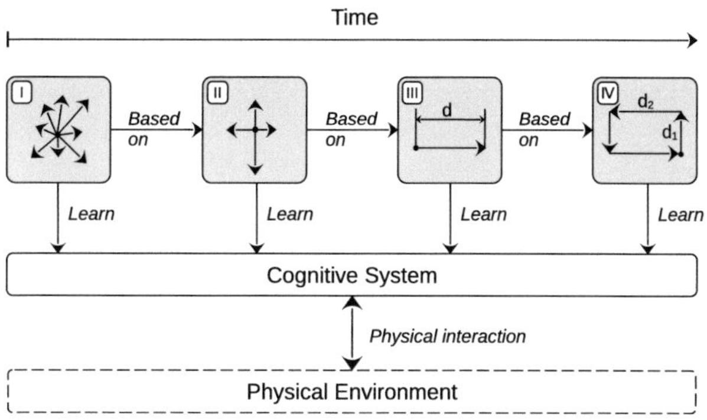

Figure 7.
The development of hierarchical knowledge over time. The motion skill acquisition starts with the fundamental question (I)"Am I able to move?", followed by (II) "Am I able to turn forward/backward/left/right?", (III) "Am I able to move a specific length/angle?" and (IV) "Am I able to follow a specific rectangular path?". While SAM draws associations directly from actions performed in the physical environment to answer these questions.

movement patterns. Turning left/right may be caused by a two-wheeled robot turning one wheel forward while the other wheel is moving backward, where forward/backward patterns may result from driving both actuators simultaneously. Hence, the system learns general natural language-based motion patterns. At the next level (III) these rules are used to learn a specific distance, say 1 cm, and angle, say 10°. Further building on this, more complex movements are learned at level (IV), which in turn consist of a series of specific movements. For example, for a rectangle with lengths of 3 cm and 2 cm, the following sequence of commands would be constructed: three times straight 1 cm, then 9 times left with 10°, two times straight with 1 cm, and so on until the rectangle is closed. Following this hierarchical knowledge acquisition approach, we can significantly limit the search space and thus bootstrap the learning performance $I - III$.

3.5.2 Basic motions

For an atomic, basic motion, we refer to the basic kinematic and dynamic properties of a system, where kinematics describes the relationship between coordinates in motion space. Dynamics correlates the torque and force in each joint (wheels of the robot) with the acceleration of the joint and the velocity over time. When the wheels touch the ground, these forces act indirectly on the overall system and thus cause it to move. With the aid of the kinematic properties, inferences about this resulting motion can be drawn. Motion control for mobile robots is extensively covered in the literature. To navigate accurately, kinematic or dynamic models are used to generate accurate motion commands, considering all effects, including the resulting tracking error [15–19]. We are aware of the challenges of designing or even learning motion controls that lead to accurate robot movements. Thus, our work demonstrates the possibility of a generic approach to learning movements with general knowledge, even if the movements are still subject to certain errors. We will address minimizing this error by following the same general approach in future work.

However, based on universal laws of physics, we derive atomic base motions, illustrated in **Figure 8**. The robot's position is represented by a vector with a pair of numerical coordinates $x(t)$ and $y(t)$ from the cares coordinate system and orientation $\psi(t)$. The robot is indirectly set in motion with constant acceleration by applying an

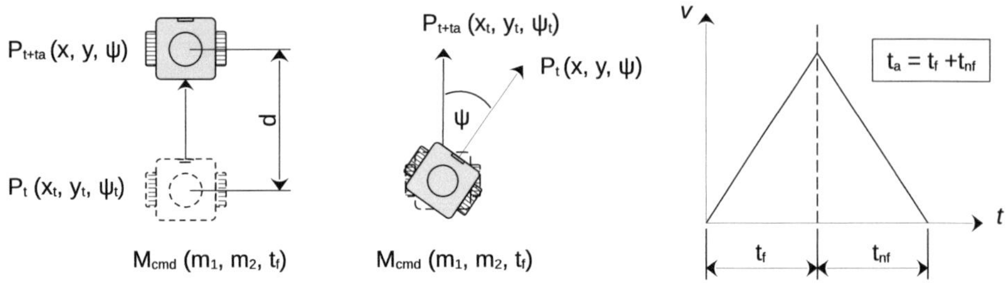

Figure 8.
Basic commands and observations for atomic motions. The left image shows a linear motion, while the middle shows an angular turn. The illustration on the right shows the respective action times and the velocity progression over time.

arbitrary torque m_1 and m_2 of the robot's actuators for a specific time t_f. As soon as this torque is removed, the system brakes with the same constant negative acceleration until it stops after some time t_{nf}. Thus, the atomic motion time is expressed by the total action time of $t_a = t_f + t_{nf}$, as depicted on the right in **Figure 8**. The resulting spatial movement (distance d and yaw angle ψ) for the respective actuator torques is determined by the change in position over time t_a using an inverse kinematic reasoner. Using RL, we search for the best actions (actuator torques and action time t_f) for a given spatial position change. This applies to all atomic actions, while more complex movements are simply composed of a series of atomic actions.

3.5.3 Kinematic reasoner

KnowRob [1] provides a kinematic reasoner, which we adapt to our SAM's needs. It derives motion-specific competence knowledge based on general kinematic laws and is utilized during hierarchical knowledge acquisition level $II - IV$. We distinguish two types of motion knowledge, (i) basic movement patterns and (ii) specific motion distances. To reason about (i), we define the following logical rules:

```
is_basic_linear_motion_pattern(X0, Y0, YAW0,
X1, Y1, YAW1, Distance): -.
DX is X1 - X0,
DY is Y1 - Y0,
Angle is wrap(YAW0, YAW1),
Distance is sqrt((DX*DX) + (DY*DY)),
Distance! = 0.0, abs(Angle) == 0.0.
is_basic_angular_motion_pattern(X0, Y0, YAW0,
X1, Y1, YAW1, Angle): -.
DX is X1 - X0,
DY is Y1 - Y0,
Angle is wrap(YAW0, YAW1),
Distance is sqrt((DX*DX) + (DY*DY)),
Distance == 0.0, Angle!= 0.0.
```

A basic linear motion pattern is detected when the robot's angle does not change during an action, but the distance does. Further, we can restrict it to a *forward* motion pattern if the position change has a positive value and *backward* if it is negative.

For an angular movement, the same rules apply. To detect angular motion patterns (*left*, *right*), we use an angle wrap function that calculates the angle moved and the direction of rotation.

For (ii), we define the following reasoner to argue about specific distances and angles used in level *III* of hierarchical knowledge acquisition.

```
is_spatial_motion(X0, Y0, YAW0, X1, Y1, YAW1,
Distance, Angle): -.
DX is X1 - X0,
DY is Y1 - Y0,
Angle is wrap(YAW1, YAW0),
Distance is sqrt(((DX*DX) + (DY*DY))) .
```

These rules represent a general knowledge of the kinematic properties of a two-dimensional system, where we argue that SAM can be easily extended to three-dimensional systems by adding appropriate kinematic reasoners.

3.5.4 Skill specific learning methods

A specific competence is explored, learned, and exploited using appropriate skill-specific learning methods. Since SAM primarily focuses on acquiring skills that lead to physical actions, the respective atomic motion commands have to be learned in real-time by interacting with the environment. An appropriate learning procedure is required, whereas reinforcement learning methods achieve good results in this domain. The method dates back to the early 1990s when Q-learning was already used to learn specific, mostly robotic, tasks. However, many works solve various tasks with RL, whereby these are primarily designed in a context-specific, goal-directed manner and without explicit general prior knowledge, which significantly limits the learning of complex skills. We attempt to overcome this with our approach by using generally formulated prior knowledge to learn skill-specific, in our case, atomic motion commands.

In the following, we introduce our RL-based approach, which we extend with KR&R methods. In RL, an agent interacts with its environment over periods of discrete time steps t. An action a_t is taken following a policy π based on the observed state s_t and the reward r_t, as shown in **Figure 9**. The main difference from traditional RL methods is that we use KR&R to infer the reward and the state. More specifically, the kinematic reasoner is applied to argue with the general kinematic knowledge about the newly observed state s_{t+1}, which in turn is defined as the distance and angle traveled during a time step t. Where the reward r_{t+1} is computed with an RL Reward Reasoner, following Eq. (2) and Eq. (3).

We chose a model-free approach for the specific RL algorithm based on a simple Q-Table RL method [20] for resource reasons. Keeping the required resources low seems to be the most intuitive first step for tackling our long-term vision, where all relevant parts of SAM, including RL, are hosted on a resource-constrained system. Q-Learning is a value-based method, where the Q-value is computed from the action-sate value function (Eq. (1)). It seeks to find the optimal Q-value for pairs of states and admissible actions. During exploration, the agent computes and stores them in a Table (Q-Table), where the Q-value indirectly represents the optimal policy π. Once the agent performs exploitation, it simply selects its actions from the Q-Table. This method performs well in systems with limited resources since it scales with the size of

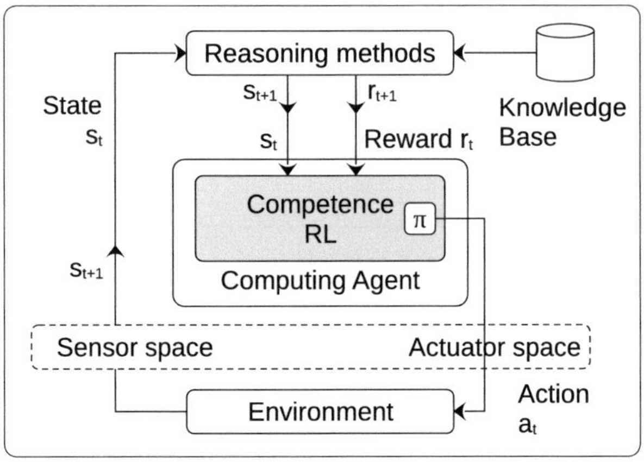

Figure 9.
RL flow with an extension of using KR&R to compute to state and reward using general knowledge.

the Q-Table in terms of resources. One significant challenge is to define the search space well, which directly affects the size of the table. We address this with our hierarchical learning approach, which constrains the respective search space quite well and thus achieves good results with Q-learning.

The Q-value function is defined according to the Bellman equation and notated as:

$$Q^{\pi}(s_t, a_t) = Q(s_t, a_t) + \alpha[r(s_t, a_t) + \gamma max Q(s_{t+1}, a_{t+1})] \quad (1)$$

Where α is the learning rate, and γ is the discount rate for the expected future reward. The action space consists of the motor force m_1, m_2, and the applied time t_f. The state space is represented by the distance covered d_{t_a} and the angle ψ_{t_a} as well as the time required t_a.

The design of the reward function is formulated to learn specific basic motion distances and angles, while the angular and linear motion skills are learned separately and denoted as:

$$r_{linear}(s_t, a_t) = 100 * (1 - abs(d_{target} - d_{t_a})) \quad (2)$$

$$r_{angular}(s_t, a_t) = 100 * \left(1 - abs\left(\psi_{target} - \psi_{t_a}\right)\right) \quad (3)$$

In principle, they each reflect a simple assumption: the closer a performed basic movement is to the desired distance, the higher the reward for that action. Thus, these rewards can also be considered a piece of specific general knowledge and are assessed by the RL Reward Reasoner. The resulting extended Q-Learning algorithm (Algorithm 1) follows a traditional flow, where the reward r_t and the state s_{t+1} are computed by the Kinematic- and RL Reward Reasoners. Their computation time is essential for systems that learn from the physical environment in real-time. In particular, these decisions must be made in a specific period, especially for tasks requiring a time-dependent control cycle, e.g., the robot is in motion and must receive its commands in time to navigate accurately. In the current work, we have solved this problem by using atomic motions that result in the robot being stationary, eliminating the time-dependent requirements during KR&R. In future work, we will investigate these considerations on real hardware that learns various skills from its environment in real-time.

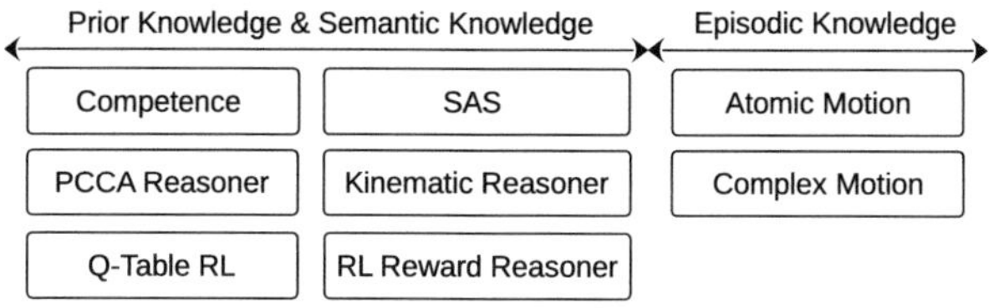

Figure 10.
A set of prior knowledge, including semantic knowledge, about competencies, SAS, PCCA, kinematic properties, and skill-specific learning methods (dashed lines) to acquire motion-related episodic memory.

Algorithm 1: Q-Learning with KR&R interactions.

Init $Q - Table$ with random data.
Observe initial state s_1.
for episode = 1, N **do.**
Select an random action action $(a_t|\pi)$.
Execute a_t.
Observe new state $s_{t+1} = Kinematic\_Reasoner()$.
Observe reward $r_t = RL\_Reward\_Reasoner()$.
Calculate Q-value Q^{π}.
Update $Q - Table$.
end for.

3.5.5 Prior knowledge and episodic memory

As discussed in Section 3.1.2, we propose to keep episodic memory flat and small and to store only highly relevant information. Further, we seek a set of general prior knowledge (semantic knowledge and general methods) that needs to be provided to learn and exhibit sufficient episodic memory for a given skill. We argue that these two facts are essential to consider for use in systems with limited resources. The following outlines how this might be addressed specifically in the case of SAM.

Figure 10 illustrates a set of prior assumptions, including semantic knowledge and general methods for acquiring motion-related episodic knowledge. The KR&R part might be provided by an edge device during the acquisition phase, while for the motion-specific learning, we deliberately propose Q-Table RL that requires few resources and thus can be hosted directly on the tiny millirobot. Further, we memorize only the motion commands learned by the RL with a timestamp and *fitness* to continuously evaluate their performance. In the case of SAM, this amount of episodic memory is sufficient to develop and improve motion skills over time. With this hybrid system flow and the conscious design of a set of generic prior and episodic knowledge, we argue that movement skills can be learned and used even on a system with limited resources. While these are general considerations, we will specifically address this subject on real hardware to consider all implications and requirements in future work.

4. Experiments

To evaluate the proposed SAM, we base our experiments on a simulation of a millirobot. Based on ROS, we use Gaezbo as a simulation environment, a Python ROS node for the computing agent, and KnowRob for the knowledge base and reasoning methods. We show the development phases $I - IV$ (Section 3.5.1), starting with evaluating a principle movement possibility up to the execution of a rectangular path. Moreover, we study two different robot models, i) a two-wheeled model and ii) a four-wheeled model.

The primary experimental question is whether SAM can a) autonomously learn a motion skill based on a small set of prior knowledge, b) evaluate and continuously improve it, c) exhibit reasonably good time performance, and d) cover a generic application on various robot models.

4.1 Two-wheeled model

In the first experiments, a two-wheeled robot with dimensions of 2 cm^2 and a mass of 100 g is used. The action space of the wheels (m_1, m_1), which expects a torque, was selected with 0.01 N m to 0.3 N m. The action period (t_f) was set to 50 ms to 1000 ms and the *fitness* has a range from 0 to 100, directly computed from the reward. In the following context, the term *step* indicates a basic movement over time t_a, while an *episode* is a set of five steps.

For the initial fundamental question, *(I)"Am I able to move?"*, SAM succeeds in the very first step and computes the *learnability* to *TRUE*. This is not surprising since as long as the two-wheeled robot is in contact with the ground, it can initiate a movement. SAM then begins learning a basic movement by randomly exploring movement patterns and reasoning about them with prior kinematic knowledge. **Figure 11** illustrates the results of level *II*, in which all patterns *(forward/backward/left/right)* were successfully found in only 50 steps, taking a total of 65 s. For a model with two actuators (wheels), the search space is manageable and works relatively fast, but the performance decreases as the number of actuators increases, which we will observe with the four-wheeled model. However, this can be addressed with suitable heuristics.

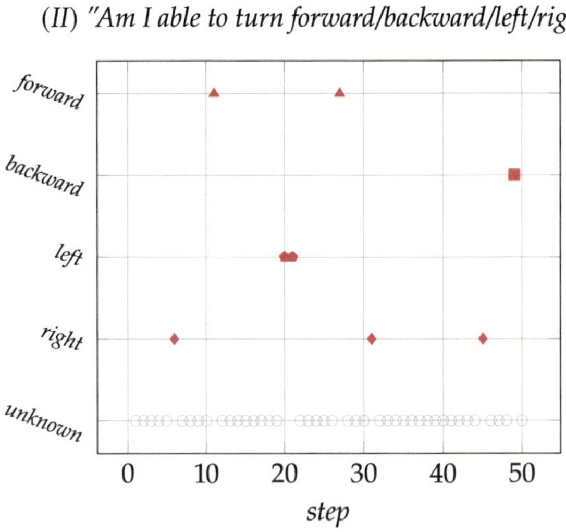

Figure 11.
Acquisition of basic motion patterns (level II - section 3.5.1). The red markers represent the respective patterns (forward = triangle, backward = square, left = pentagon, right = diamond) argued and identified in a particular step (physical interaction) with kinematic knowledge.

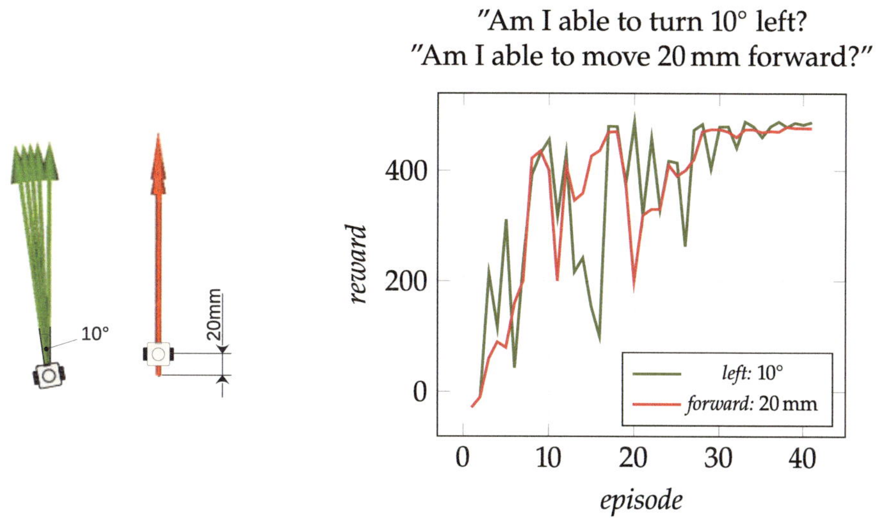

Figure 12.
Acquisition of specific motion distance and angle (level III - section 3.5.1). The left illustration shows the successful learned specific motions (green: 10 and red: 20 mm). The right image depicts the accumulated reward from the Q-table RL per episode, where an episode consists of five steps.

At the next level (*III*), these rules are used to learn a forward liner movement $a_{lin}\{fw\}\{20mm\}$ with a specific distance of 20 mm and a left turn angular movement $a_{ang}\{lt\}\{10^{\circ}\}$ with an angle of 10°. The RL Q-Table learning is applied for each motion action, where the reward (0–100) directly represents the *fitness* of each. **Figure 12** depicts the learning performance of $a_{ang}\{lt\}\{10^{\circ}\}$ (green) and $a_{lin}\{fw\}\{20mm\}$ (red). The reward settles at episode 35, with the total time of the 50 episodes averaging 4 min 30 s. The respective learned motion commands are:

$$\begin{aligned} &a_{ang}\{lt\}\{10^{\circ}\}: \ m_1 = -0.15\text{Nm}, \ m_2 = 0.15\text{Nm}, \ t_f = 355.71\text{ms}, \ fitness = 83, \ and. \\ &a_{lin}\{fw\}\{20\text{mm}\}: \ m_1 = 0.26\text{Nm}, \ m_2 = 0.26\text{Nm}, \ t_f = 450.71\text{ms}, \ fitness - 80. \end{aligned} \tag{4}$$

In the first attempt, we achieve relatively good results in an early phase, after only a few minutes. This is promising for use in resource-constrained systems, as it meets the resource requirements for migration mentioned earlier. The acquired competence knowledge is further used in level *IV* to accomplish a more complex skill. **Figure 13** shows the execution of a complex motion(rectangular path), where the continuous improvement of the respective motion commands is investigated. The blue rectangular path shows the first attempt using the learned angular and linear motions ($a_{ang}\{lt\}: fitness = 83$ and $a_{lin}\{fw\}: fitness = 80$). Clearly visible, the fitness is not yet sufficiently developed to follow a reasonably good rectangular path. In the following, SAM tries to improve those (using PCCA) over several iterations until a sufficient fitness ($threshold = 99$) is learned. After about 60 min it has improved its capabilities and successfully navigates the red rectangular path significantly better than the green (after 25 min) one. When performing complex actions, it is also clearly visible (red path) the effects of a small movement error, which accumulates over further steps. This is due to the non-consideration of the actual respective error of action. In this work, we consciously accept this fact, but we will attempt to reduce it in a general way in further work.

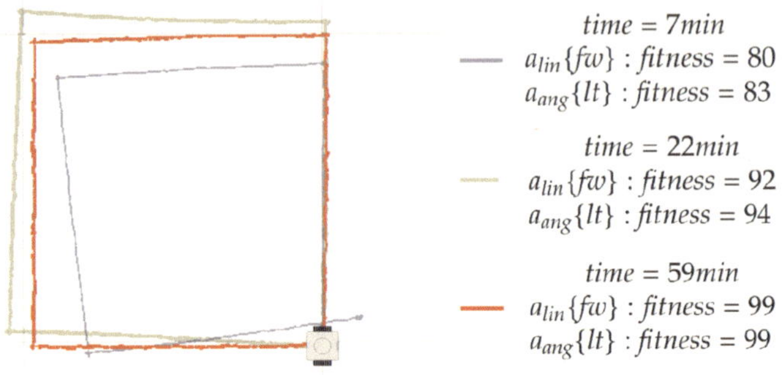

Figure 13.
The exploitation of complex motion following a rectangle path (level IV - section 3.5.1). The respective color shows the development (improvement) over time, starting with the blue path (accomplished with the commands learned from the first few attempts), followed by the green, and finally the red, representing the best movement competence.

4.2 Four-wheeled model

Further, we extend our experiments to a four-wheeled robot model as a first step to verify the general applicability of SAM. The basic assumptions and implementations of the robot model remain the same, except for two additional wheels. Due to these two further actuators, the search space increases, which leads to significant differences in the learning phase (level *I*) of the motion patterns (*forward/backward/left/right*), depicted in **Figure 14**. Unlike the two-wheeled model, SAM requires significantly more time, i.e., 200 steps (four-wheeled model) instead of the previous 50 steps (two-wheeled model). However, this was expected and will become even more complex with other systems, such as drones (acting in three-dimensional space). Once this phase is overcome, SAM can achieve the same good RL

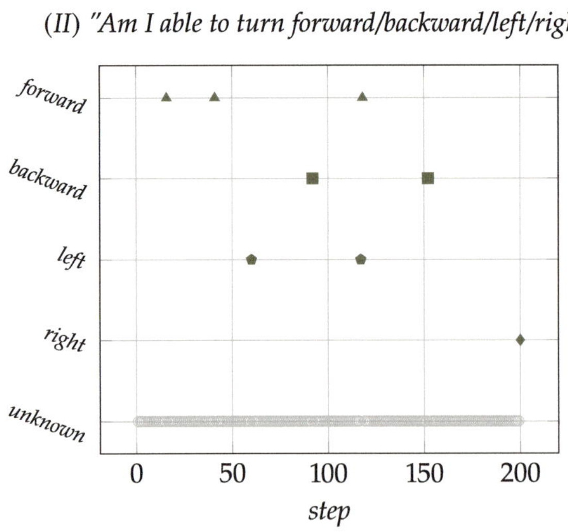

Figure 14.
Acquisition of basic motion patterns (level II - section 3.5.1) of the four-wheeled robot. The green markers represent the respective patterns (forward = triangle, backward = square, left = pentagon, right = diamond) argued and identified in a particular step (physical interaction) with kinematic knowledge. In contrast to the two-step model, SAM requires significantly more time, i.e. 200 steps instead of the previous 50 steps (two-wheeled model).

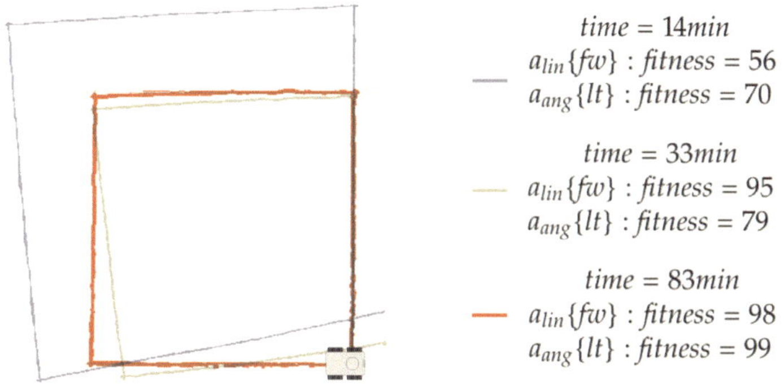

Figure 15.
Exploitation of complex motion following a rectangle path (level IV - section 3.5.1) or the four-wheeled robot. The respective color shows the development (improvement) over time, starting with the blue path (accomplished with the commands learned from the first few attempts), followed by the green, and finally the red, representing the best movement competence. No significant differences to the two-wheeled model can be identified in this skill level due to the hierarchical learning approach.

results (level *III*) with the four-wheeled as with the two-wheeled model. We did not experience any significant difference in terms of learning performance.

The reason for this is the hierarchical learning approach, where the level above is abstracted from the level below in terms of performance. This gives us confidence that SAM is well suited for generalization. The last image of our experiments shows the development of the rectangular path by the four-wheeled robot with SAM, which was successfully mastered in 50 min (see **Figure 15**).

In summary, we have demonstrated with our experiments that SAM can learn autonomously complex motion skills based on a small set of prior knowledge and can further develop them with reasonable good time performance. We showed the first step for a generic application to various robot models by demonstrating the different wheel-based models.

5. Conclusions

In this article, we introduced SAM, which starts with a set of general prior knowledge and appropriate methods to autonomously acquire and develop specific complex skills. It combines methods of KR&R with methods of learning from the physical environment and aims to be applied in resource-constrained systems. We proposed a cognitive behavior (PCCA), which enables the continuous acquisition of skills, their evaluation, and the further development and adaptation of already learned skills. To this end, we modeled generic competencies using ontologies and formulated SAS based on elementary physical quantities to build a generic interface to the physical environment. Specifically, we demonstrated SAM based on motion skills learned through a general knowledge of kinematics laws and geometry. Further, we applied hierarchical knowledge acquisition with RL to acquire basic and more complex movements. We argue that this approach is general because the only assumptions we make are the laws of kinematics and geometry, the availability of and access to sensors and actuators, and the availability of a database describing the skills to be learned. Based on this generic knowledge, we demonstrated the acquisition of basic motion and a complex movement where the robot successfully moved along a rectangular path. To prove the generic approach, we evaluated it through experiments with a two-wheeled and four-wheeled millirobot. Where the acquisition performance in terms of resources delivers promising results for further deployment of the method in resource-constrained systems.

Thus, in the first step, we have demonstrated a cognitive system that develops more complex behaviors with a set of general prior knowledge and appropriate methods to function in arbitrary environments. In this work, we still assume that the robot knows the meaning of the actuators and sensors, although these do not necessarily have to be present a priori. In the next step, we want to remove this assumption. There is promising work in automatic semantic knowledge acquisition for sensor and actuator data that could help address this problem in a meaningful manner, which we will investigate further. Moreover, we will continue to develop an even more general approach, where an exhilarating challenge in this context could be the applicability of our method in a three-dimensional system. In addition, there are still limitations to the use of KR&R methods in resource-constrained systems, which we discussed in this work. Another medium-term goal is to study SAM in resource-constrained systems. Therefore, we will specifically address the transition to a real resource-constrained system in the form of a millirobot. In summary, our first results indicate that the use of SAM has an advantage for generic applicability, and we will continue to try to advance this approach.

Abbreviations

RL	Reinforcement Learning
SAM	Skill Acquisition Method
PCCA	Playful Continuous Competence Acquisition
SAS	Sensor and Actuator Space
KR&R	Knowledge Representation and Reasoning
SOMA	Socio-physical Model of Activities
IMU	Inertial Measurement Unit

Author details

Markus D. Kobelrausch*† and Axel Jantsch†
Technical University of Vienna, Vienna, Austria

*Address all correspondence to: markus.kobelrausch@tuwien.ac.at

† These authors contributed equally.

References

[1] Tenorth M, Beetz M. KNOWROB — knowledge processing for autonomous personal robots. In: 2009 IEEE/RSJ International Conference on Intelligent Robots and Systems. 2009. pp. 4261-4266. DOI: 10.1109/IROS.2009.5354602

[2] Nilsson NJ. Shakey the Robot. Technical Report 323, AI Center, SRI International. CA, USA: Menlo Park; 1984

[3] Tenorth M, Beetz M. KnowRob: A knowledge processing infrastructure for cognition-enabled robots. The International Journal of Robotics Research. 2013;**32**(5):566-590

[4] Tenorth M, Beetz M. Representations for robot knowledge in the KnowRob framework. Artificial Intelligence. 2017; **247**:151-169

[5] Beetz M, Tenorth M, Winkler J. Open-EASE. In: 2015 IEEE International Conference on Robotics and Automation (ICRA). 2015. pp. 1983-1990. DOI: 10.1109/ICRA.2015.7139458

[6] Beetz M, Beßler D, Haidu A, Pomarlan M, Bozcuoğlu AK, Bartels G. Know Rob 2.0 — A 2nd Generation Knowledge Processing Framework for Cognition-Enabled Robotic Agents. In: 2018 IEEE International Conference on Robotics and Automation (ICRA). 2018. pp. 512-519. DOI: 10.1109/ICRA.2018.8460964

[7] Beßler D, Porzel R, Pomarlan M, Vyas A, Höffner S, Beetz M, et al. Foundations of the socio-physical model of activities (SOMA) for autonomous robotic agents. Frontiers in Artificial Intelligence and Applications. Volume 344. arXiv preprint arXiv: 2011.11972. 202

[8] Riazuelo L, Tenorth M, et al. RoboEarth semantic mapping: A cloud enabled knowledge-based approach. IEEE Transactions on Automation Science and Engineering. 2015;**12**(2):432-443

[9] Wyatt JL, Aydemir A, et al. Self-understanding and self-extension: A systems and representational approach. IEEE Transactions on Autonomous Mental Development. 2010;**2**(4): 282-303

[10] Sheth A, Henson C, Sahoo SS. Semantic sensor web. IEEE Internet Computing. 2008;**12**(4):78-83

[11] Ganz F, Barnaghi P, Carrez F. Automated semantic knowledge acquisition from sensor data. IEEE Systems Journal. 2016;**10**(3):1214-1225 Conference Name: IEEE Systems Journal

[12] Moore T, Stouch D. A generalized extended Kalman filter implementation for the robot operating system. In: Menegatti E, Michael N, Berns K, Yamaguchi H, editors. Intelligent Autonomous Systems 13. Vol. 302. Cham: Springer International Publishing; 2016. pp. 335-348

[13] Alatise MB, Hancke GP. Pose estimation of a Mobile robot based on fusion of IMU data and vision data using an extended Kalman filter. Sensors. 2017; **17**(10):2164

[14] Durrant-Whyte H, Bailey T. Simultaneous localization and mapping: Part I. IEEE Robotics Automation Magazine. 2006;**13**(2):99-110 Conference Name: IEEE Robotics Automation Magazine

[15] Fierro R, Lewis FL. Control of a nonholomic mobile robot: Backstepping

kinematics into dynamics. Journal of Robotic Systems. 1997;**14**(3):149-163

[16] Morin P, Samson C. Motion control of wheeled Mobile robots. In: Siciliano B, Khatib O, editors. Springer Handbook of Robotics. Berlin, Heidelberg: Springer; 2008. pp. 799-826

[17] Martins FN, Celeste WC, Carelli R, Sarcinelli-Filho M, Bastos-Filho TF. An adaptive dynamic controller for autonomous mobile robot trajectory tracking. Control Engineering Practice. 2008;**16**(11):1354-1363

[18] Das T, Kar IN. Design and implementation of an adaptive fuzzy logic-based controller for wheeled mobile robots. IEEE Transactions on Control Systems Technology. 2006; **14**(3):501-510 Conference Name: IEEE Transactions on Control Systems Technology

[19] Antonini P, Ippoliti G, Longhi S. Learning control of mobile robots using a multiprocessor system. Control Engineering Practice. 2006;**14**(11): 1279-1295

[20] Christopher JCH. Watkins and Peter Dayan. Q-learning. Machine Learning. 1992;**8**(3):279-292

Chapter 4

An Episodic-Procedural Semantic Memory Model for Continuous Topological Sensorimotor Map Building

Wei Hong Chin, Naoyuki Kubota and Chu Kiong Loo

Abstract

For humans to understand the world around them, learning and memory are two cognitive processes of the human brain that are deeply connected. Memory allows information to retain and forms an experiences reservoir. Computational models replicating those memory attributes can lead to the practical use of robots in everyday human living environments. However, constantly acquiring environmental information in real-world, dynamic environments has remained a challenge for many years. This article proposes an episodic-procedure semantic memory model to continuously generate topological sensorimotor maps for robot navigation. The proposed model consists of two memory networks: i) episodic-procedural memory network (EPMN) and ii) semantic memory network (SMN). The EPMN comprises an Incremental Recurrent Kernel Machines (I-RKM) that clusters incoming input vectors as nodes and learns the activation patterns of the nodes for spatiotemporal encoding. The SMN then takes neuronal activity trajectories from the EPMN and task-relevant signals to update the SMN and produce more compact representations of episodic experience. Thus, both memory networks prevent catastrophic forgetting by constantly generating nodes when the network meets new inputs or updating node weights when the incoming input is similar to previously learned knowledge. In addition, idle or outlier nodes will be removed to preserve memory space.

Keywords: episodic memory, semantic memory, sensorimotor map, topological map, robot navigation

1. Introduction

One of the essential features of common living locomotive organisms is their capability to traverse their daily environment with life-critical tasks. For example, rats can learn to visit or avoid places of food that they have visited, and squirrels are excellent at rediscovering places of food that they have previously hidden. Many animals escape to a previously visited shelter if they are undergoing an urgent threat, such as a bear that escapes to a cave for hibernation to preserve energy during the winter season. A specious hypothesis is that living organisms should have a cognitive mechanism to represent their environment as a collection of important regions, such as nest locations and food places. When necessary, they can recall these regions and utilize their relations to perform navigation tasks [1].

The capability of an autonomous mobile robot to represent its environment as a spatial map and to determine its position concurrently has been widely analyzed in the robotics society. The process is termed SLAM (Simultaneous Location and Mapping), and several state-of-the-art have been introduced that works remarkably well [2, 3]. Another research area is to generate a topological map that maps the environment's structure. Robots can plan trajectories and navigate to target locations using topological graphs. However, the sophistication of maps increases exponentially with the length of the robot's journey in most current graph-based approaches [4]. If new nodes and edges are added to the map continuously, the requirement for processing time and memory storage increases over time, stopping applications from long-term mapping. As a result, methods for controlling the scale of the topological map are critical in functional robotic applications that require continuous exploration in environments [5–7].

Biological methods do not appear to experience enormously from the deficiencies mentioned above in artificial navigation [1]. For example, rats can explore, search, and travel in large and dynamic environments for a long time. They can adapt to the environment changes quickly, for instance, searching new ways if a previously visited route is unavailable or choosing potential shortcuts when new access spots are available. Therefore, several computer goal-oriented navigation systems were introduced to partially emulate how the brain could represent space and apply these representations for navigation tasks. Memory is a fundamental perspective for the acquisition of experience. Memory is essential for the understanding, learning, and cognition of the interactions of robots in complex environments [8]. Episodic memory is a kind of memory that retains human experiences in a particular and conscious way.

This article proposes an episodic-procedural semantic memory model for topological sensorimotor map construction. The robot can use the generated topological sensorimotor map to perform indoor navigation. The following are our contributions to this study: i) The proposed model can learn multiple sensory information to generate the topological sensorimotor map incrementally; ii) Because of the nature of episodic memory attributes, the robot can perform goal navigation with path-planning algorithms; iii) The semantic memory layer can serve as a medium for humans to interact with a robot to perform navigation tasks, and iv) The proposed method continuously updates the generated topological map (can expand or shrink) to maintain the size of the map based on the environment without the need for human interference.

2. Related works

Many practical approaches to solving the SLAM problem have been introduced in robot mapping. Lu and Milios [9] were the first to use a pose graph to implement global map optimization. The robot's poses are represented as vertices in a graph, and the spatial boundaries between poses are represented as edges in a graph. The map's scale proliferates in this traditional graph-based approach as the robot discovers new regions. Consequently, there is a rapid rise in the need for storage and computing resources. When direct linear solvers are applied, the traditional graph-based solution has quadratic memory consumption growth with the number of variables in the worst case. Efforts to increase the performance of graph-based mapping algorithms are underway. In standard equations, the sparsity structure of the matrix is used to allow quick linear online solvers. Many SLAM libraries are available to deal with this dilemma with tens of thousands of variables in just a few seconds, such as g2o [10], and RTAB-Map [11]. Memory usage increases linearly with the number of variables, even using iterative linear solvers. Returning to the exact location many times complicates the case. This strategy becomes less effective as more vertices and edges are applied to the same spatial area. For the time being, there are only a few works that attempt to answer how to store a map for long-term exploration. Consequently, achieving a long-term mapping solution [5] that can control, or at the very least restrict, the size of the map is essential.

Vertex and edge sparsification, which trades map precision for memory and computational power, is one of the most effective techniques to reduce the map's complexity. To avoid redundant vertices and insert informative measurements to the map, an information-based compact pose SLAM algorithm was proposed in an information-theoretic fashion [12]. In pose global optimization, an information-based criterion was adopted to determine the laser scans should be marginalized, maintaining the sparsity of laser-based 2D pose maps. To obtain a light blanket based on the Markov blanket of a boundary vertex, the generic linear constraint criteria [13] and nonlinear graph sparsification were proposed [14].

Another approach was introduced that focused on solving the traditional pose graph's temporal scalability [15]. This approach eliminates the addition of redundant vertices and edges before the graph's global optimization. This approach has been demonstrated in indoor areas using a binocular visual SLAM framework, and it is an effective solution for medium-scale environments such as houses and factories. The idea of neighborhood area and scene integration is introduced [7] to achieve sparsification of the cognitive map without adding unnecessary vertices and edges to the cognitive map.

One of the biologically-inspired proposed methods is RatSLAM [16, 17]. The approach represents the environment as a set of pose cells, and each pose cell is linked to a view cell. RatSLAM was successfully implemented in small and large environments for spatial mapping, but the framework does not handle target-oriented navigation. Erdem and Hasselmo [18] proposed a biologically inspired computational model for goal-oriented navigation. In this model, the environment is represented as several grid cells with different scales and spacing and gradually converge into one place cell. The model gradually recruits new place cells to encode the autonomous agent's current location when the agent meets a notable location during exploration. Each place cell has a reward cell, and the lateral weight of the connection between two reward cells is equivalent to the time between the autonomous agent's successive visits to the reward cells. With the lateral connections, autonomous agents can navigate to the goal location from its starting location. However, the methods mentioned above focus on emulating place cells and grid cells for spatial map building.

Humans seem to accommodate themselves better in complex environments and recall past experiences to perform tasks simultaneously generate new experiences and skills. These significant behaviors usually develop from experiences that rely on learning. Likewise, the assumption is that experience also implies for robots [19]. Thus, the learned experiences can be integrated into a spatial map so that robots can freely observe and navigate in any environment. Current methods rely on the RatSLAM concept, such as BatSLAM [20] using sonar sensing, which has been developed. Tang et al. [21] included an episodic memory module in navigational tasks to process contextual information. The approach is designed for maze-controlled situations, but its effectiveness in open spaces such as corridors, offices, and homes is still unknown.

3. Proposed method

The proposed model consists of two hierarchical memory networks: i) episodic-procedural memory and ii) semantic memory. New nodes (experiences) are generated in each memory network as new sensory information is obtained. Topology links are generated to connect nodes and store robot behaviors. These connections provide the robot with procedural knowledge so that an action can be taken to proceed from one circumstance to another. The episodic-procedural network is an Incremental Recurrent Kernel Machines (I-RKM) which incrementally cluster incoming input data as nodes in an unsupervised fashion. The I-RKM is the Infinite Echo State Network extension [22, 23]. Each node in the network further encodes an activation value used for spatiotemporal learning. The semantic memory network is hierarchically connected to the episodic memory network. It is also another I-RKM that receives bottom-up inputs from the episodic memory network and top-down signals such as labels or signs for generating representations that contain semantic knowledge on a larger timescale. The mechanism of neural operation in the semantic memory network is similar to the episodic procedural memory network with an additional requirement to create a new node. In this network, node learning happens as the network correctly predicts the class label of the classified input sequence from the episodic memory network through the learning process. A new node will only be created if the incorrect network class label. This criterion is also the additional element that modulates nodes update. In particular, each semantic node preserves information over time sequences higher than episodic nodes due to the hierarchical learning of input data.

The episodic network serves as a novelty detector in the robot navigation mission. Each node in the network represents a group of related input features and creates new nodes if the incoming input features do not fit into any network nodes. Nodes in the episodic network also encode the robot's location for localization purposes. In addition, each link encodes a robot's action, such as turning angle and moving speed, to serve as procedural information that allows the robot to perform a sequence of actions and travel from one place to another. Each node encodes the semantic meaning of human operator cues in the semantic network. Semantic definition marks the explored space with various names, such as a hallway, room, or kitchen, to provide a medium for human-robot interaction. If no external sensory information is available, the episodic procedural memory network performs an action-oriented internal simulation through the playback of node sequences and actions encoded in their links to consolidate knowledge (memory) and mitigate catastrophic forgetting. Each node in the SMN represents a region of the environment. The robot utilizes this information to change its moving behaviors, such as wall following, obstacle avoidance, or fast travel. **Figure 1** shows the overview of the proposed method.

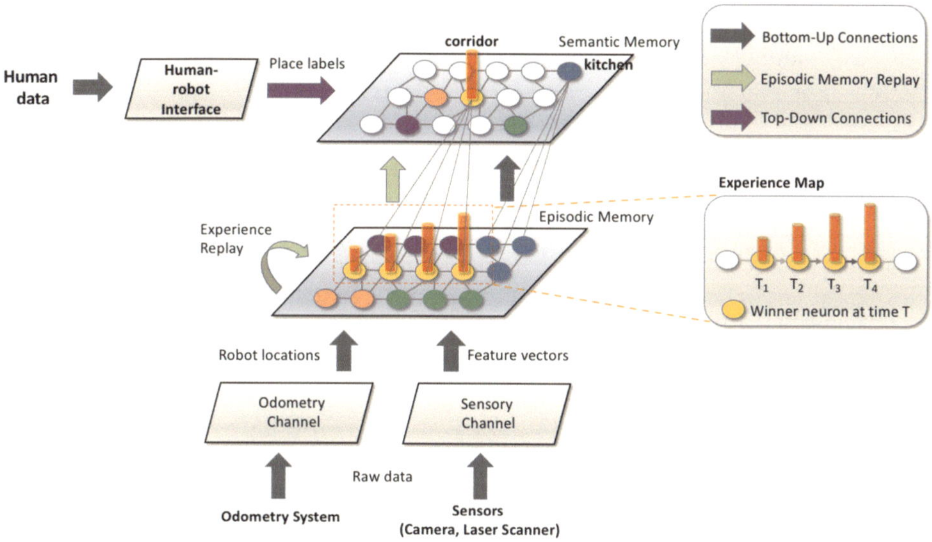

Figure 1.
The overview of the proposed method which consists of two memory networks: The episodic procedural memory network and semantic memory network. The episodic procedural memory network clusters incoming sensory input as nodes progressively and learns fine-grained spatiotemporal correlations between them. The semantic memory network adjusts the amount of architectural flexibility based on task-relevant inputs to build a topological semantic map with more compact episodic representations.

3.1 Echo state network

Echo State Networks [24] can be considered large, randomly recurring neural networks with a single sequential, trained readout layer. The network computes a wide range of non-linear, spatial–temporal mappings of input data. The reservoir can be seen as a spatial–temporal kernel in which the mapping of a high-dimensional space is explicitly computed. Hermans et al. [22] proposed a Recurrent Kernel Machines (RKM) that extends Echo State Networks' idea to infinite-sized recurrent neural networks (RNNs). The proposed method is regarded as recursive kernels. When a RNN with internal weights W, input weights V, and an internal state s receives an input x_t at time t, it produces the following output:

$$y_t = h(Vx_t + Ws_t) \tag{1}$$

where h is the product of the activation function (for example, the hyperbolic tangent) and the projection function. A recursive method's core idea is that Eq. (1) can be represented as follows:

$$h(Ws_t + Vx_t) = h\left([W|V]\begin{bmatrix} s_t \\ x_t \end{bmatrix}\right) \tag{2}$$

It's a function of the input's concatenation with the prior internal state. The same reasoning can be applied to kernel functions, with the base function inputs consisting of a concatenation of the current input and the prior recursive mapping:

$$\phi(x_t, \phi(x_{t-1}, \phi(\dots))) = \phi([x_t|\phi(x_{t-1}|\phi(\dots))]) \tag{3}$$

Hermans et al. [22] has shown that recursive variations of kernels with $k(x,x') = f(\|x - x'\|^2)$ and $k(x,x') = f(x \cdot x')$ form can be derived using this structure as a reference. For example, the recursive-SE kernel has the form:

$$\kappa_t^{SE}(x,x') = \exp\left(-\frac{\|x_t - x_t'\|^2}{2l^2}\right)\exp\left(\frac{\kappa_t^{SE}(x,x') - 1}{\sigma_p^2}\right) \tag{4}$$

We propose a computational model called Incremental Recurrent Kernel Machines (I-RKM) for continuously creating topological maps based on characteristics of RKM. The EPMN and the SMN are two hierarchical memory levels in the proposed method. The I-RKM is described in-depth in the following sections.

3.2 Episodic procedural memory network (EPMN)

An I-RKM constitutes the EPMN. In reaction to input vectors, the network dynamically grows or contracts by adding or removing nodes. To encode node relationships, edges will be created to connect nodes. The I-RKM notations are tabulated in **Table 1**.

Based on the sensory input, the network first generates two recurrent nodes. Each node in the network is comprised a weight vector w_j. For further learning, the network uses the Eqs. (5) and (6) to identify the node that best fits the current sensory input $x(t)$. Eq. (6) creates the Infinite Echo State which is identical to Eq. (4).

$$b = \arg\min\left(T_j(t)\right) \tag{5}$$

$$T_j(t) = \exp\left[-\frac{\|x^c(t) - w_j^c\|^2}{2\sigma_i^2}\right]\exp\left[\frac{T_j(t-1) - 1}{\sigma^2}\right] \tag{6}$$

Notation	Definition
$T_j(t)$	Activation value of node j at t
$\kappa(t)$	Recursive kernel at t
$w_b(t-1)$	Best matching node weights at $t-1$
r_j	Regularity counter of node j
γ_j	Contributing factor of node j
τ_j, λ	Decay factors for regularity counter
σ_i^2, σ^2	Kernel width
ρ	Learning threshold
$P_{(m,n)}$	Temporal connection between node m and n
V	Associative matrix for labeling
b	Index of best matching node

Table 1.
The notations of I-RKM.

Following that, the activation value of the best matching node (BMN) J is determined as follows:

$$a_b(t) = \exp(-T_b) \tag{7}$$

If the activation value $a_b(t)$ is smaller than a predefined threshold a_T, the condition is fulfilled. A new node N is added to the network with the following weights:

$$w_N = 0.5 \cdot (x(t) + w_b) \tag{8}$$

To connect the winning node b and the second BMN, a new link is established. If $a_b(t)$ is greater than a_T, the winning node b can represent the input $x(t)$. As a result, the winning node b and its neighbor nodes n are updated as follows in response to input $x(t)$:

$$w_{j(\text{new})} = \gamma_j \cdot r_j \cdot \left(x(t) - w_{j(\text{old})}\right) \tag{9}$$

If no connection exists between the BMN $a_b(t)$ and the second-best matching node, a new connection will be made to connect them. Each edge has an age counter that grows by one with each iteration. The age of the link between the best and second-best matching nodes is reset to zero. Nodes with no connections and a habituation counter larger than the preset value will be removed from the network, as will connections with an age greater than the preset threshold. In addition, each episodic node has a regularity counter $r_j \in [0, 1]$ that indicates the strength of its firing over time. The value of the newly formed episodic node is $r_j = 1$. Using the following equation, the regularity value of the BMN and its adjacent nodes decreases with each iteration:

$$\Delta r_j = \tau_j \cdot \lambda \cdot (1 - r_j) - \tau_j \tag{10}$$

As a result, the significance of the node's regularity can be associated with the relevance or importance of the information stored in the node. Regularity values for nodes that have been often activated in response to learning inputs are presented in the regularity Eq. (10). If the link exceeds the threshold, isolated nodes will be removed from its network. Due to the nature of the network, the topological network expands during the robot's journey in the robot navigation mission. However, nodes generated at the start of the journey are eliminated from its network. Thus, we have introduced a new criterion of node removal [25] with the following equation:

$$v = \mu(H) + \sigma(H) \tag{11}$$

where H is a vector representation of the network's regularity, μ is the mean function, and σ is the standard deviation. Nodes with regularity values more than the threshold will be removed.

Only if $b_J(t) < \rho_b$ and $r_J < \rho_r$ can a new episodic node be added to the network. If the activation and regularity thresholds are met, the episodic nodes will be updated via Eq. (9). In the EPMN, a set of events constitutes an episode, which retains distinct historical occurrences and episodes that are linked to one another. To learn recurrent node activation patterns in the network, we incorporate temporal connections. Temporal connections represent the sequence of activated nodes throughout the learning stage.

A temporal connection between the two consecutively activated nodes will be enhanced by 1 for each learning iteration. When the BMN b is activated at time t and then again at time $t-1$, the temporal relationship between them is reinforced as follows:

$$P^{new}_{(b(t),b(t-1))} = P^{old}_{(b(t),b(t-1))} + 1 \tag{12}$$

For each recurrent node m, the next node g from the encoded time series can be obtained by selecting the largest value of P as shown below:

$$g = \arg\max P_{(m,n)} \tag{13}$$

where n are the neighbors of m. As a result, the recurrent node activation sequence can be reestablished without the need for any further input data.

3.3 Semantic memory network (SMN)

The semantic memory layer is linked to the episodic memory layer hierarchically. It is made up of an I-RKM that obtains bottom-up inputs from the episodic memory layer and top-down inputs such as labels or tags to develop representations that incorporate semantic information over a more extended period. By delivering signals from the top-down signals, semantic information could be retrieved.

The mechanism of neural activity in the SMN is similar to that of the EPMN, with the requirement for the creation of new nodes. Node learning happens in this layer when the network accurately predicts the class label of the labeled input sequence from the EPMN during the learning process. If the class label is incorrect, a new node will be added. This additional criterion influences the rate at which the nodes update. Furthermore, due to the hierarchical learning of incoming data, each semantic node maintains knowledge through periods higher than episodic nodes. As a result, the SMN selects the winning node based on the BMN of the EPMN in the following manner:

$$b_s = \arg\min\left(T_j^{SMN}(t)\right) \tag{14}$$

$$T_j^{SMN}(t) = \exp\left[-\frac{\|x(t)-w_j\|^2}{2\sigma_i^2}\right]\exp\left[\frac{T_j^{SMN}(t-1)-1}{\sigma^2}\right] \tag{15}$$

The selected node is either assumed to be the correct semantic node for the given sequence of episodic inputs, or it is more dominant than other semantic nodes, or both. The SMN receives input data from the EPMN, i.e., the EPMN's BMNs with regard to $x(t)$. The BMNs in the network are calculated with the Eqs. (14) and (15). Because the input is derived from bottom-up neural episodic weights, $x(t)$ is substituted by w_b^{em} for node learning.

Thus, a new semantic node is created only if the BMN b fails to satisfy three criteria: 1) $a_b^{sm}(t) < \rho_a$; 2) $r_b^{sm} < \rho_r$; and 3) BMN's label ζ_b^{sm} is not the same as the data input's label ζ (Eq. (21)). It should be noted that if the data input is not labeled, this label matching requirement in the semantic memory layer is ignored. If the winner of the semantic node b predicts the label ζ_b that is the same as the class label ζ of the input $x(t)$, the node learning process is started by the extra learning factor $\psi = 0.001$. As a result, Eq. (9) will become:

$$w_{j(\text{new})}^{\text{SMN}} = \psi \cdot \gamma_j \cdot r_j \cdot \left(w_b^{\text{EPMN}} - w_{j(\text{old})}^{\text{SMN}} \right) \tag{16}$$

The SMN learns to create more compact representations of the input labels. Data labels govern the network's stability and plasticity, with new semantic nodes addition only when the network is unable to estimate the correct data input class label.

3.4 Episodic procedural memory self-replay

To generate meaningful sequential data for memory playback, we exploit the spatiotemporal connections of nodes in the EPMN. When there is no input feed into the network, the EPMN uses its nodes as input for learning (self-replay). For example, if the winning episodic node b is activated by input data, the next temporal node can be selected by choosing the node with the largest activation value of P. For each node j, a set of nodes playback with length $K^{\text{EPMN}} + 1$ is calculated as follows:

$$U_j = \left\langle w_{u(0)}^{\text{EPMN}}, w_{u(1)}^{\text{EPMN}}, \cdots, w_{u\left(K^{\text{EPMN}}\right)}^{\text{EPMN}}, \right\rangle \tag{17}$$

$$u(i) = \arg\max P_{(j,u(i-1))} \tag{18}$$

where K^{EPMN} is the number of temporal nodes, $P(i,j)$ is the episodic temporal connection matrix, and $u(0) = j$. The temporal connection of episodic nodes stored in the network is capable of autonomously generating a series of events and replaying to the network without retaining the relations of previously received training data.

3.5 Data associative system

During the training phase, each node can be assigned a class label of l based on the input data. The L class label yields the l label. The frequency of each individual label in the network is stored in the $V(j,l)$ associative matrix for this labeling approach. This implies that each node j has a distribution counter that holds the frequency of a certain sample label. When a new node N is created and the label ζ associated with the input data $x(t)$ is specified, the matrix V is enlarged by one row and initialized with $V(N,\zeta) = 1$ and $V(N,l) = 0$. When an existing BMN b is chosen for updating, the V matrix is updated in the following manner:

$$V(b,\zeta)_{(\text{new})} = V(b,\zeta)_{(\text{old})} + \varphi^{+} \tag{19}$$

$$V(b,l)_{(\text{new})} = V(b,l)_{(\text{old})} + \varphi^{-} \tag{20}$$

Notice that φ^{+} must always less than φ^{-} and the label ζ is within the L class label. If the data label ζ does not exist in L, a new column in V is added and set to $V(b,\zeta) = 1$ and $V(b,l) = 0$. The matrix V will not be updated if there is no label associated with the given input gesture. The winning label ζ_j for a node j is calculated as follows:

$$\zeta_j = \text{label}(j) \equiv \arg\max V(j,l), \tag{21}$$

where l is label in class label L. The advantage of this labeling approach [26] is that no number of class labels must be specified in advance. Because the number of class label is uncertain, this is crucial when dealing with continuous learning in real world application.

4. Experimental setup and results

We first validate the proposed method using the COLD benchmark dataset [27, 28]. The COLD dataset is a large-scale, customizable testing environment for generally validating vision-based localization algorithms intended to perform on mobile platforms in realistic environments. A mobile robot gathers the dataset in three separate locations with different environmental conditions such as weather conditions, day or night time. It contains various formats, including RGB images, videos, and laser scans. RGB images and videos are gathered using a standard onboard camera and an omnidirectional camera. Instead of learning the image pixels, we use fixed random weights of Convolutional Neural Network (CNN) [29] for extracting visual features that sufficiently express the environment states. A simple CNN with fixed random weights, for example, can extract visual information with high classification accuracy in image classification tasks [30]. In this work, the extracted features from fixed random weights CNN and the robot's odometry data will be inputs to the EPMN, and the output of the EPMN will be the input of the SMN. Each data is fed into the memory networks sequentially without repetition for topological map building. Unlike batch learning, feeding the data sequentially to the memory networks fulfill the continuous learning criteria where data is only seen once. This criterion is crucial for robot navigation as the robot often traverses the environment continuously from one place to another. The hyperparameters for training the I-RKM in both memory networks are tabulated in **Table 2**.

Several metrics have been developed to assess the quality of a topological memory network. The total quantization error (TQE) is a popular metric, which quantifies the average distance between each data vector and its BMN. The BMU is the winning node in our case since it has the most significant match value and fulfills the vigilance parameter. The TQE measures the fitness of the generated topological map to the

Hyperparameter	Value
α^1, α^2	0.5
ρ_a	0.75
ρ_r	0.1
τ_j	0.5
λ	1.05
σ_i^2, σ^2	1.1, 1.4
γ_b	0.2
γ_n	0.001
r_e	0.001

Table 2.
Training parameters for the I-RKM of the memory networks.

robot's actual navigation route. As a result, the ideal topological map is expected to have the lowest TQE. The lower the TQE, the smaller the average distance between the BMNs and the robot's actual trajectory, indicating that the topological map is closer to the original route.

Furthermore, we evaluate the feasibility of the generated topological map using node localization accuracy. The pre-processed image dataset is transmitted to the I-RKM of both memory networks for each iteration to determine the BMN. The Euclidean distance between the BMN's encoded position and the robot's position from the dataset is used to compute the localization accuracy. Localization is accomplished if the Euclidean distance is smaller than a predefined value (0.1 m in these experiments). Because the purpose of SMN is to encode location label information, the localization accuracy is computed differently. Localization in SMN is fulfilled if the BMN's encoded location label is the same as the label from the dataset, similar to the standard classification accuracy.

4.1 Benchmark dataset results

The odometry and pre-processed image datasets were utilized as input to the I-RKM in the benchmark dataset experiment. To accomplish self-memory replay in EPMN, we continually feed the data in a mini-batch fashion (10 data per mini-batch).

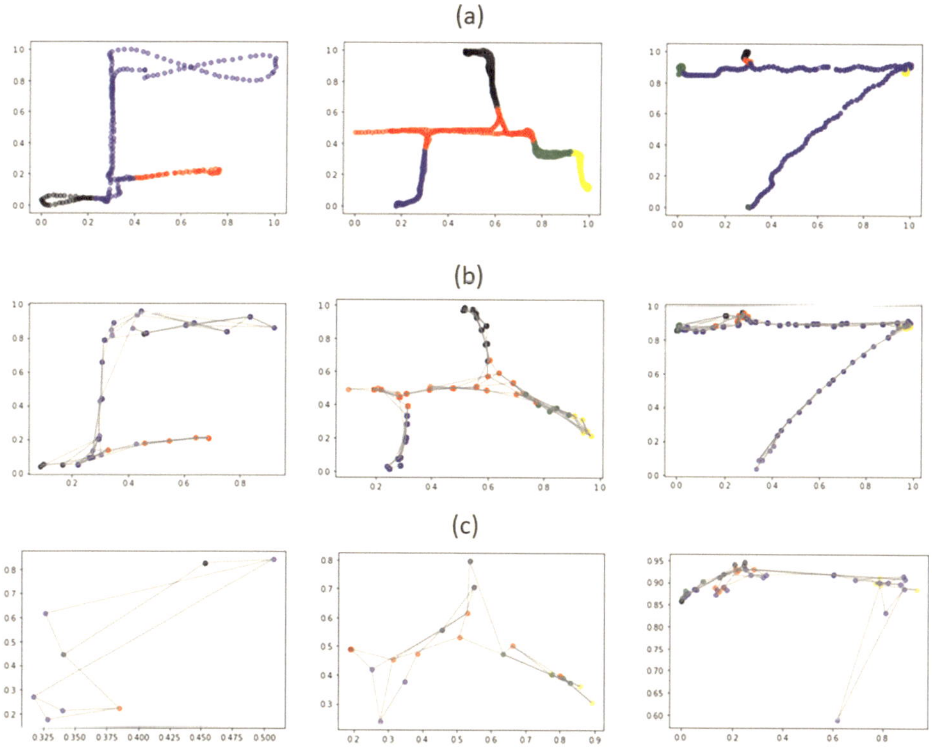

Figure 2.
Row (a) shows the robot's real path for collecting the COLD dataset: Saarbruken, Freiburg and Ljubljana (from left to right). Rows (b) and (c) illustrate the topological map of the episodic-procedural memory network and the semantic memory network, respectively. (a) robot navigation path, (b) episodic procedural memory network, (c) semantic memory network.

Then, after each mini-batch, memory self-replay was triggered. The topological map is made up of a series of nodes and edges. Different colored circles represent nodes, and each one holds the robot's coordinates (x, y), a place label, and a feature vector representing the surroundings. Links are black lines that link all nodes in the map to indicate node relationships. **Figure 2** shows the exact path taken by the robot in three different buildings with different environmental conditions and the topological maps generated by the proposed method. **Table 3** shows the TQE and localization accuracy of the topological map for each dataset. TQE and localization accuracy was found to be relatively constant across datasets. As a result, memory network learning is consistent across buildings with varying environmental conditions.

4.2 Physical robot experiment results

We validated our suggested technique further utilizing a mobile robot attached to an iPhone for image data acquisition and an Intel i5 CPU NUC PC for processing data and controlling the robot as shown in **Figure 3(a)**. The robot can traverse the surroundings autonomously, avoid obstacles, and follow walls. The robot's movement speed ranges from 0.05 to 0.5 m/s. The EPMN receives data from the iPhone and odometry to produce a topological map, whereas the SMN accepts EPMN output as input.

The experiments were carried out on the 7th floor of a university hallway, study area, and rest space that connected with one other, as shown in **Figure 3(b)**. The

Dataset	TQE (EPMN)	TQE (SMN)	Accuracy (EPMN) (%)	Accuracy (SMN) (%)
Freiburg cloudy 1	0.0283	0.2048	94.0	94.4
Freiburg cloudy 2	0.0322	0.2793	94.0	96.4
Freiburg cloudy 3	0.0204	0.2112	91.4	94.3
Freiburg sunny 1	0.0093	0.2122	94.0	95.0
Freiburg sunny 2	0.1179	0.2495	95.3	96.0
Freiburg sunny 3	0.0228	0.1466	91.5	92.0
Ljubljana cloudy 1	0.0143	0.2613	91.8	93.0
Ljubljana cloudy 2	0.0898	0.2743	84.9	85.0
Ljubljana cloudy 3	0.0047	0.3568	91.6	85.7
Ljubljana sunny 1	0.0118	0.2231	94.3	95.1
Ljubljana sunny 2	0.0480	0.2273	93.3	94.0
Ljubljana sunny 3	0.0798	0.3939	90.2	91.5
Saarbruken cloudy 1	0.0661	0.1544	91.5	92.0
Saarbruken cloudy 2	0.0020	0.1700	82.1	83.0
Saarbruken cloudy 3	0.0990	0.1410	91.7	93.1
Saarbruken night 1	0.0063	0.1292	92.1	93.6
Saarbruken night 2	0.0016	0.1678	89.5	90.2
Saarbruken night 3	0.1075	0.1899	85.0	86.9

Table 3.
The TQE and localization accuracy of the topological map that generated by the memory networks using COLD datasets.

purpose of experimenting with such environmental settings is to confirm that our proposed technique can work in a natural environment with moderately varying environmental factors. We instructed the robot to explore the experimental site, beginning in the study area and traveling to the rest area through the hallway, then returning to the start point. I-RKM continually learns from incoming sensory data in both memory networks and builds the topological map. After the first traverse, self-memory replay is triggered before the next traverse begins. The robot explored the surroundings with various movement behaviors depending on the location. For example, in the study area, the robot is set to obstacle avoidance mode since the environment is crowded with moving people and objects. Because the hallway is a straight path, the movement behavior is altered to the wall following and fast-speed mode when the robot enters the hallway. We repeated the experiment ten times. The metrics evaluation is identical to the benchmark dataset experiments (**Figure 4**). **Figure 5(a)** and **(b)** show the TQE and localization accuracy of the memory networks respectively.

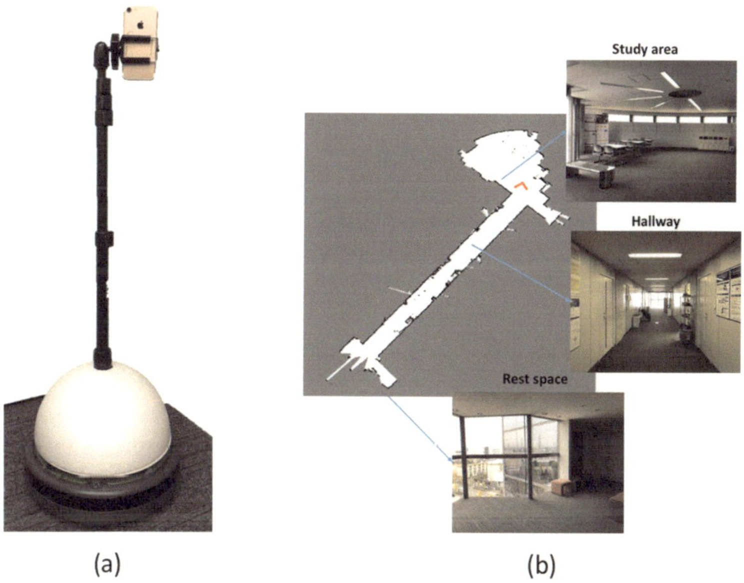

Figure 3.
(a) Physical robot equipped with an iPhone. (b) The experimental environment.

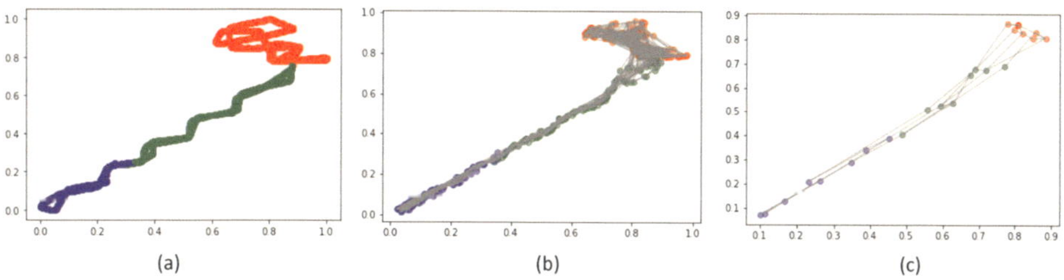

Figure 4.
(a) Robot navigation path; (b) topological map generated by the EPMN; (c) topological map generated by the SMN.

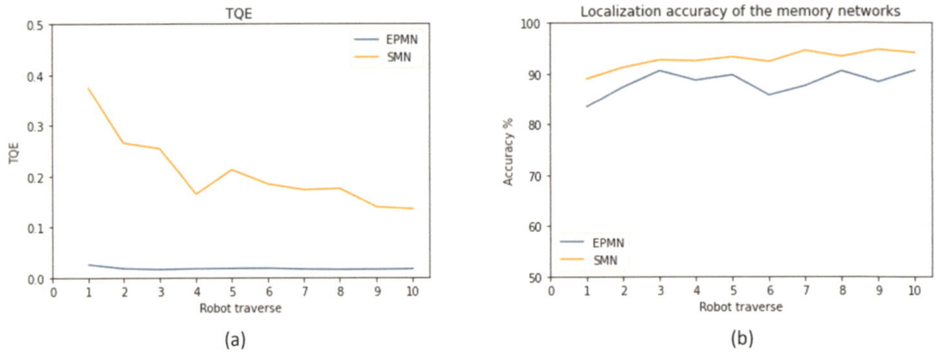

Figure 5.
(a) TQE of the topological map generated by the memory networks; (b) localization accuracy of the memory networks.

5. Discussion

We have shown that the memory networks can generate topological maps with benchmark datasets and physical robot experiments. Topological maps are built up from nodes that encode specific sensory information, providing flexibility and maintainability for robot navigation. New nodes are constantly added to the memory networks during environment learning, or existing nodes are updated. Edges link new nodes to existing nodes and can be used to guide navigation activities. Each node represents a region of the world, and it will be selected for learning if it corresponds to the robot's current sensory data. This property demonstrates that I-RKM retains previously learned knowledge and creates a topological map based on the robot's traverse path. According to the experiment results, all of the topological maps generated by I-RKM are almost identical to the actual robot path.

Because of the nature of memory network learning, the EPMN generates more nodes than the SMN. Because the SMN will use the EPMN output to generate the topological map, the SMN will learn the more sparse category representation. EPMN's topological map can be utilized for robot localization and navigation. The topological nodes connection allows the robot to navigate from one location to another. The topological map in SMN is sparser than in EPMN, and the TQE is higher than in the EPMN. However, the topological map of the SMN can be utilized for place classification tasks.

The proposed memory network training takes odometry data into account and visual measures. As a result, memory networks can distinguish areas with relatively similar visual sensory input, overcoming the difficulties of online detection and recognition of topological nodes. According to the node matching and localization findings, the robot failed to locate itself during navigation on several occasions because of a sudden change in the environment, resulting in no topological nodes matching with these sensor data. This issue can be solved by adjusting the vigilance parameter. The higher the value of the vigilance parameter, the more sensitive the memory networks are to changing environmental conditions and vice versa.

6. Conclusion

We presented Incremental-Recurrent Kernel Machines that mimic human episodic-procedural semantic memory and can progressively learn the spatiotemporal connection of sensory input from camera and odometry to build a topological map. I-RKM in both memory networks autonomously updates the topological map by expanding or shrinking its episodic memory structure. Furthermore, I-RKM consolidates the spatial map through self-episodic memory replay, eliminating the requirement for external sensory inputs. I-RKM has been validated through benchmark datasets and physical robot implementation. In the future, we will combine I-RKM with a path planning algorithm to use the topological map's structure for goal-directed navigation. In addition, we plan to leverage the edges connection between nodes by encoding traverse information on the edges. The robot can navigate from one place to another autonomously that solely depends on memory with little or no human intervention. Finally, we will improve and test I-RKM's performance in more challenging and larger environments.

Acknowledgements

This work was partially supported by JST [Moonshot R&D] [Grant Number JPMJMS2034].

Author details

Wei Hong Chin[1*], Naoyuki Kubota[1] and Chu Kiong Loo[2]

1 Tokyo Metropolitan University, Tokyo, Japan

2 University of Malaya, Kuala Lumpur, Malaysia

*Address all correspondence to: weihong@tmu.ac.jp

References

[1] O'Keefe J, Nadel L. The hippocampus as a cognitive map. Behavioral and Brain Sciences. 1979;**2**(4):487-494. DOI: 10.1017/S0140525X00063949

[2] Brooks R. A robust layered control system for a mobile robot. IEEE Journal on Robotics and Automation. 1986;**2**(1): 14-21. DOI: 10.1109/JRA.1986.1087032

[3] Thrun S. Learning metric-topological maps for indoor mobile robot navigation. Artificial Intelligence. 1998;**99**(1):21-71. DOI: 10.1016/S0004-3702(97)00078-7

[4] Henrik K, Cyrill S. Information-theoretic compression of pose graphs for laser-based SLAM. The International Journal of Robotics Research. 2012;**31**(11): 1219-1230. DOI: 10.1177/02783 64912455072

[5] Cadena C, Carlone L, Carrillo H, Latif Y, Scaramuzza D, Neira J, et al. Past, present, and future of simultaneous localization and mapping: Toward the robust-perception age. IEEE Transactions on Robotics. 2016;**32**(6): 1309-1332. DOI: 10.1109/TRO.2016. 2624754

[6] Chin WH, Loo CK, Toda Y, Kubota N. An Odometry-free approach for simultaneous localization and online hybrid map building. Frontiers in Robotics and AI. 2016;**3**:68-77. DOI: 10.3389/frobt.2016.00068

[7] Zeng T, Si B. A brain-inspired compact cognitive mapping system. Cognitive Neurodynamics. 2021;**15**:91-101. DOI: 10.1007/s11571-020-09621-6

[8] Buzsáki G, Moser EI. Memory, navigation and theta rhythm in the hippocampal-entorhinal system. Nature Neuroscience. 2013;**16**(2):130-138. DOI: 10.1038/nn.3304

[9] Lu F, Milios E. Globally consistent range scan alignment for environment mapping. Autonomous Robots. 1997;**4**: 333-349. DOI: 10.1023/A:1008854305733

[10] Kümmerle R, Grisetti G, Strasdat H, Konolige K, Burgard W. G2o: A general framework for graph optimization. In: Proceedings of the IEEE International Conference on Robotics and Automation (ICRA 2011); 9–13 May 2011; Shanghai, China. IEEE; 2011. pp. 3607-3613

[11] Labbé M, Michaud F. RTAB-map as an open-source lidar and visual simultaneous localization and mapping library for large-scale and long-term online operation. Journal of Field Robotics. 2019;**35**:416-446. DOI: 10.1002/ rob.21831

[12] Ila V, Porta JM, Andrade-Cetto J. Information-based compact pose SLAM. IEEE Transactions on Robotics. 2010; **26**(1):78-93. DOI: 10.1109/TRO.2009. 2034435

[13] Carlevaris-Bianco N, Eustice RM. Generic factor-based node marginalization and edge sparsification for pose-graph SLAM. In: Proceedings of the IEEE International Conference on Robotics and Automation (ICRA 2013); 6–10 May 2013; Karlsruhe, Germany. IEEE; 2013. pp. 5748-5755

[14] Mazuran M, Burgard W, Tipaldi GD. Nonlinear factor recovery for long-term SLAM. The International Journal of Robotics Research. 2016;**35**(1–3):50-72. DOI: 10.1177/0278364915581629

[15] Johannsson H, Kaess M, Fallon M, Leonard JJ. Temporally scalable visual SLAM using a reduced pose graph. In: Proceedings of the IEEE International Conference on Robotics and Automation

(ICRA 2013); 6–10 May 2013; Karlsruhe, Germany. IEEE; 2013. pp. 54-61

[16] David B, Scott H, Janet W, Gordon W, Peter C, Milford M. OpenRatSLAM: An open source brain-based SLAM system. Autonomous Robots. 2013; **34**(3):149-176. DOI: 10.1007/s10514-012-9317-9

[17] Milford M, Jacobson A, Chen Z, Wyeth G. RatSLAM: Using Models of Rodent Hippocampus for Robot Navigation and beyond. Robotics Research: The 16th International Symposium ISRR2016, Springer Tracts in Advanced Robotics. pp. 467-485. DOI: 10.1007/978-3-319-28872-727

[18] Erdem UM, Hasselmo ME. A biologically inspired hierarchical goal directed navigation model. Journal of Physiology Paris. 2014;**108**(1):28-37. DOI: 10.1016/j.jphysparis.2013.07.002

[19] Endo Y. Anticipatory robot control for a partially observable environment using episodic memories. In: Proceedings of the IEEE International Conference on Robotics and Automation (ICRA 2008); 19–23 May 2008; Pasadena, CA, USA. IEEE; 2008. pp. 2852-2859

[20] Steckel J, Peremans H. BatSLAM: Simultaneous localization and mapping using biomimetic sonar. PLoS One. 2013; **8**(1):e54076. DOI: 10.1371/journal.pone.0054076

[21] Tang H, Yan R, Tan KC. Cognitive navigation by neuro-inspired localization, mapping, and episodic memory. IEEE Transactions on Cognitive and Developmental Systems. 2018;**10**(3):751-761. DOI: 10.1109/TCDS.2017.2776965

[22] Hermans M, Schrauwen B. Recurrent kernel machines: Computing with infinite Echo state networks. Neural Computation. 2012;**24**(1):104-133. DOI: 10.1162/NECOa00200

[23] Soh H, Demiris Y. Spatio-temporal learning with the online finite and infinite Echo-state Gaussian processes. IEEE Transactions on Neural Networks and Learning Systems. 2015;**26**(3): 522-536. DOI: 10.1109/TNNLS.2014.2316291

[24] Jaeger H. The "Echo State" Approach to Analysing and Training Recurrent Neural Networks-with an Erratum Note. Bonn, Germany: German National Research Center for Information Technology GMD Technical Report; 2010. p. 148

[25] Liew WS, Loo CK, Gryshchuk V, Weber C, Wermter S. Effect of pruning on catastrophic forgetting in growing dual memory networks. In: Proceedings of the International Joint Conference on Neural Networks (IJCNN 2019); 14–19 July 2019; Budapest, Hungary. IEEE; 2019. pp. 1-8

[26] Parisi GI, Tani J, Weber C, Wermter S. Lifelong learning of human actions with deep neural network self-organization. Neural Networks. 2017;**96**: 137-149. DOI: 10.1016/j.neunet.2017.09.001

[27] Wang X, Zhao Y, Pourpanah F. Recent advances in deep learning. International Journal of Machine Learning and Cybernetics. 2020;**1**:747-750. DOI: 10.1007/s13042-020-01096-5

[28] Pronobis A, Caputo B. COLD: The CoSy localization database. The International Journal of Robotics Research. 2009;**28**(5):588-594. DOI: 10.1177/0278364909103912

[29] Saxe AM, Koh PW, Chen Z, Bhand M, Suresh B, Ng AY. On random weights

and unsupervised feature learning. In: Proceedings of the International Conference on International Conference on Machine Learning (ICML 2011); June 28–2 July 2011; Madison, WI, USA. Omnipress; 2011. pp. 1089-1096

[30] Tong Z, Tanaka G. Reservoir computing with untrained convolutional neural networks for image recognition. In: Proceedings of the International Conference on Pattern Recognition (ICPR 2018); 20–24 August 2018; Beijing, China. IEEE; 2018. pp. 1289-1294

Chapter 5

A Robotics-Based Machine Learning Approach for Fall Detection of People

Teddy Ordoñez Nuñez, Raimundo Celeste Ghizoni Teive and Alejandro Rafael Garcia Ramirez

Abstract

For a person when carrying out household chores or even when walking on the streets, there is a risk of falling. This risk increases throughout the years due to the natural aging process. In this work, a bibliographic review was performed to find related papers who discussed different techniques for fall classification. The aim of this study was to develop two ML models: an SVM and a k-NN model, to classify the fall. An accelerometer, gyroscope, and magnetometer located on the waists of 15 volunteers are the application sensors. The extracted features were the mean, standard deviation, and range for each sensor. The best accuracy obtained was 93.89%, a sensitivity of 85.10%, and a specificity of 96.99%. All results were obtained by simulations, by using the test set separated in the first stage of the implementation. So, a shortcoming is the fact that the ML models were not tested with a hardware implementation. In future works, the models can be embedded into a microcontroller and classify data in real time.

Keywords: k-NN, SVM, inertial measurement unit, elderly, falls, wearables

1. Introduction

As the years go by, bodies become weaker and thus give up their physical health. It can lead to new problems and challenges for the elderly because there comes a time when they need to be more cautious, and not everyone can be that way. And it is in this context that falls among the elderly are becoming more and more frequent. Falls among them have more consequences than a scrape on their bodies. People over 60 are gradually becoming more vulnerable to falls [1].

Falls among the elderly happen suddenly and are very frequent. According to Ref. [2], about 30% of people over 65 years old suffer a fall at least once a year, increasing to 50% when they are over 80. Falls are a problem of worldwide interest, which brings consequences to people and governments due to the heavy investment to recover its citizens. Therefore, researchers are always looking for solutions to improve people's quality of life.

Since 1991 the authors in Ref. [3] began studies to use wearable sensors to solve this problem. Other works in this field were in Refs. [4, 5], which proposed a protocol for evaluating the performance of any developed system.

Usually, those devices are at high end and challenging for a consumer with a low income to acquire because of the costs. The two most popular ways to detect falls are video [6] and measuring signals from an accelerometer placed on the body [7]. There are vast possibilities for integrating these devices with machine learning (ML) techniques to correctly classify data received from video streaming or sensors placed on the body.

2. Falls

"Fall detection involves complex pattern recognition, which tends to vary according to each individual who suffers a fall" [8]. According to Ref. [1], falls can be defined as "an event that results in a person unintentionally stopping their activities on the ground, floor, or a lower level." Falls can also be defined as "falling to the floor or some other lower level as a consequence of receiving a violent blow, loss of consciousness, paralyzes such as a stroke or a seizure of epilepsy" [9]. Approximate 684,000 fatal falls occur each year, with 80% of these fatalities concentrated in low- and middle-income countries [1].

According to Ref. [9], most falls happen in the sagittal and coronal planes, as shown in **Figure 1**. These names are related to the human body and its anatomy. It is worth noting that when a fall occurs with the loss of consciousness, as described in Ref. [9], that is when the body suffers more. It is due to the lack of absorption of impact since the body falls directly to the ground. When a fall happens, the person is conscious can absorb the impact by stretching their arms to protect themselves if they fall forwards.

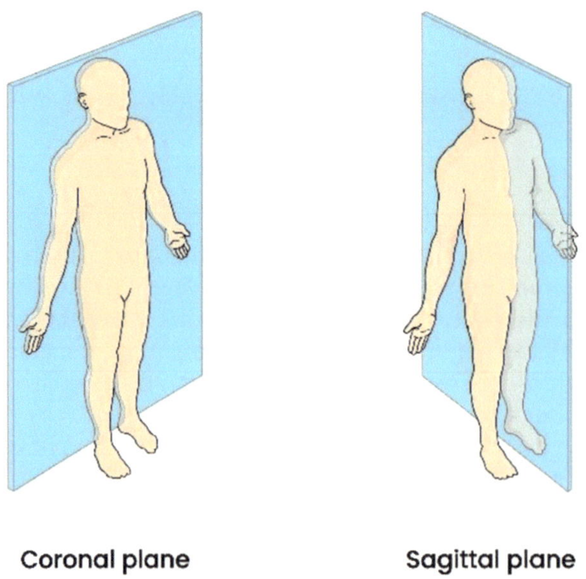

Figure 1.
Sagittal and coronal planes of the human body.

Serious injuries include traumatic brain injury, concussion, hemorrhages, and cuts [6]. In Brazil, the Sistema Unico de Saude (SUS) spends more than R$51 million annually treating various fractures because of falls [6]. According to Ref. [10], approximately one in three adults who live in their homes suffers a fall annually. And of those adults, about half of them will experience falls more frequently. According to Ref. [1], numerous factors can influence a person to suffer a fall, and among the most prominent are age, gender, and health.

2.1 Factors who contribute to falls

According to Ref. [9], the age factor is not enough to describe the risk of a person falling; therefore, a person is more likely to fall depending on several other factors. It is worth noting the risk of an elder suffering a fall is higher due to the inherent aging process. The factors that contribute to the event of a fall can be separated into two categories: intrinsic and extrinsic [6, 9].

Intrinsic factors are those that depend on the person, such as medication use, low muscle mass percentage, dizziness, and lightheadedness [6]. Among these factors, Ref. [9] also includes osteoporosis, Parkinson's, dementia and cognitive problems, inadequate lifestyle, vision problems, chronic diseases, and previous falls. An inadequate lifestyle is directly linked to a sedentary lifestyle since physical activity helps to strengthen muscles [6].

Extrinsic factors are external to the individual [6]. Among them are slippery floors, stairs, inadequate footwear, crowded places, low light conditions, and damaged sidewalks [1]. Poor condition sidewalks represent a worrying problem in Brazil, based on a study conducted by Ref. [11]. They found that the average score attributed to sidewalks in several cities, on a scale of 1 to 10, is 3.40. A good score for the quality of sidewalks would be 8.0 [11].

2.2 Consequences of falling

There are several consequences because of a fall. Falls as an outcome of accidents are one of the reasons for hospital admissions and the leading cause of death among people over 65 years [9]. Among the types of consequences, Refs. [6, 9] emphasize physical and psychological damage, and in addition, Ref. [9] also mentions financial losses. Serious injuries are related to physical consequences. The most common minor wounds are bruises and scrapes [9]. There are many serious injuries, such as concussions, bleeding, skull trauma, and fractures [6].

According to Ref. [6], the most common consequence among the psychological type is fear of suffering new falls, but still Ref. [9] also mentions the lower quality of life, loss of independence, low self-esteem, and limited abilities. The economic implications are just as important as others because of the medical expenses. Among these expenses are rehabilitation therapies, medical examinations, hospitalizations, and the purchase of medical equipment [9]. Due to such arguments, it is a must to prevent falls. **Figure 2** shows an example of a fall registered by the three sensors considered in this work. For every sensor, there are three individual graphs.

In **Figure 2**, one can observe a graph created using the accelerometer, gyroscope, and magnetometer readings while simulating a forward fall. This is a simulation of a fall caused by fainting or syncope forwards. These three sensors are located in the person's waist.

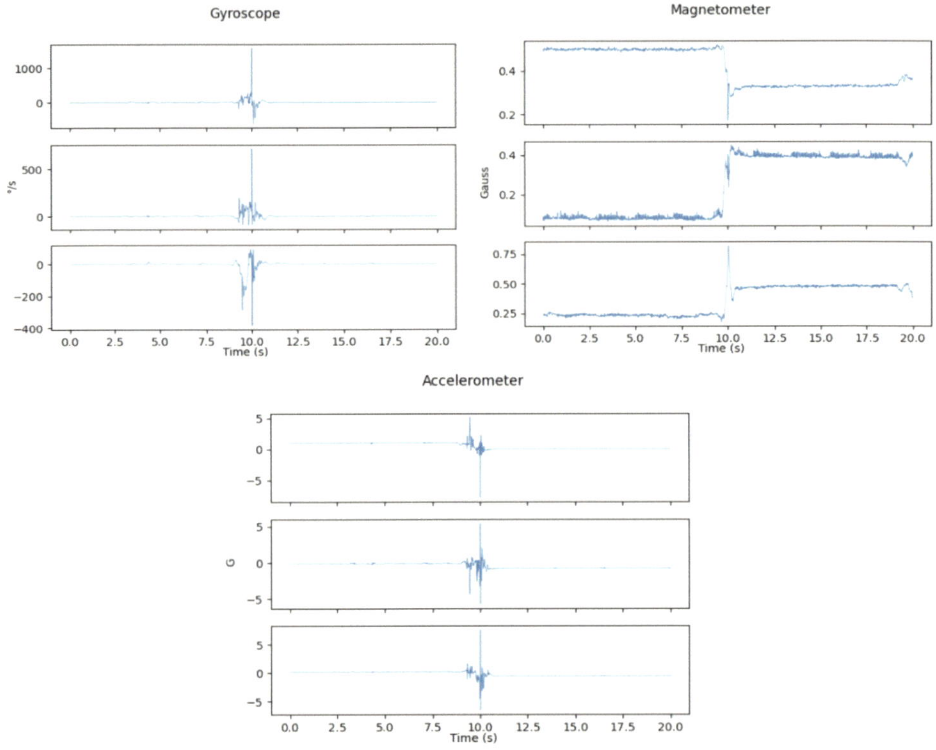

Figure 2.
A fall registered by the application sensors.

In this example, the volunteer stands up until the ninth second. When this mark is reached the person falls forwards, simulating a consciousness loss. At this moment, there is an abrupt change in the sensor's readings, and the accelerometer's value reached its peak at ±5 g. There was an impact, and towards around the 10th second, the volunteer hit the ground and remained in this position (this scenario did not consider recovery after impact).

2.3 Related works

Bibliographic research was carried out through the Univali Integrated Library System (SIBIUN), which performs a search in the Univali collection, CAPES Portal, EBSCO, Biblioteca A, Saraiva, Vlex, Scielo Livros, Scielo Periodicals, and Open Access Directories. The search strings "Machine Learning" AND "Fall Classification" were used, yielding 184 results. After reading the abstracts, four relevant studies were selected.

In Ref. [12], three sensors collected data from an accelerometer, a gyroscope, and a magnetometer. This group of sensors were placed in five places on the volunteers' body, such as on their head, chest, waist, wrist, and legs. The authors used six different ML techniques, including k-nearest neighbor (k-NN), support vector machines (SVM), least square method, Bayesian decision making, dynamic time warping, and artificial neural networks. Overall, the work scored optimal results, with an accuracy of 99.91%, a sensitivity of 100%, and a specificity of 99.79% [12]. The best accuracy was achieved by the k-NN algorithm, with 99.1% [12].

In Ref. [13], was carried out a similar work by using the same sensors as previously cited. However, authors placed the sensors only on the waist of the volunteers since the human body's center of mass is located there. To perform the signal classification were used three stages of a fall. These stages are impact, post-impact, and posture. The proposed solution is based on a threshold comparison to identify each one of the stages. It is worth noting that in Ref. [13], SVM was used to extract thresholds for each phase. With this proposed solution, the result was 100% accuracy, sensitivity, and specificity for the classification [13].

The work in Ref. [6] differs from the related studies. In particular, the authors used an accelerometer and a gyroscope embedded in a smartphone to capture the sensors signals and classify them. A belt was used to secure the smartphone to the volunteer's waist. Like [12], this study used the idle time. After detecting the inactivity time, data were classified using a decision tree and a threshold classifier and verified the actual orientation of the device. If all verifications are true, a fall is notified. The system in Ref. [6] achieved an accuracy of 93.25%, a sensitivity of 95.45%, and a specificity of 87.65%.

The most recent work is Ref. [14]. The authors also used all three sensors. They created the dataset FallAllD, which is available to the academic community. The volunteers used the set of sensors on three parts of their body: the chest, wrist, and waist. The authors explore four different ML techniques to classify falls: k-NN, SVM, random forest classifier, and convolutional neural network. Although all the three sensors collect data, only the accelerometer readings were used to train the ML models, looking for a simplified operation. The authors found an accuracy of 89.70%, a sensitivity of 95.06%, and a specificity of 95.20% when applying the k-NN technique. The implementation of the SVM technique with a quadratic kernel achieved an accuracy of 85.86%.

In Ref. [15], the authors demonstrate techniques not only to reliably detect a fall but also to automatically classify the type. Fifteen volunteers simulate four different types of falls-left and right lateral, forward trips, and backward slips—while wearing mobile phones. They applied five machine learning classifiers to a large time-series feature set to detect falls. Support vector machines and regularized logistic regression were able to identify a fall with 98% accuracy and classify the type of fall with 99% accuracy.

In Ref. [16], the authors present a comprehensive literature review on various ML-based classifications in fall detection. The authors identify the main problems in threshold-based classification from existing works and find the need for an efficient ML-based classification technique to accurately identify the fall. In addition, the shortcomings associated with the ML-based techniques for future research and other problems, such as data preprocessing and data dimensionality reduction techniques, are investigated. They concluded that ML-based techniques are far superior to threshold-based techniques.

Table 1 shows the comparison between the related works.

3. Development

In this work, the Python programming language was used. Besides the built-in library, we used other embedded resources to manipulate the data samples, that is, to create the ML models and to generate the confusion matrices. In addition, the Pandas' library was used to manipulate the data. This library is popular among Data Scientists

Characteristics	[12]	[13]	[8]	[14]	This work
Dataset	*Erciyes University*	*DOFDA*	MobiFall, MobiFall2, & own	FallAllD, Sisfall, & UMA-Fall	*FallAllD*
Number of volunteers	10 men & 7 women	6 men & 2 women	4 youngsters & 4 elders	8 men & 7 women	8 men & 7 women (simulation)
Sensors*	A, G & M	A, G & M	A & G	A, G, M & B	A, G & M
Groups of sensors	6	1	1	3	1
ML algorithms	K-NN, LSM, SVM, BDM, DTW, ANN	Threshold based	Binary tree & threshold	*k*-NN, SVM, LSTM & other	SVM & *k*-NN

**A = accelerometer, G = gyroscope, M = magnetometer, and B = barometer.*

Table 1.
Comparison between related works and algorithms.

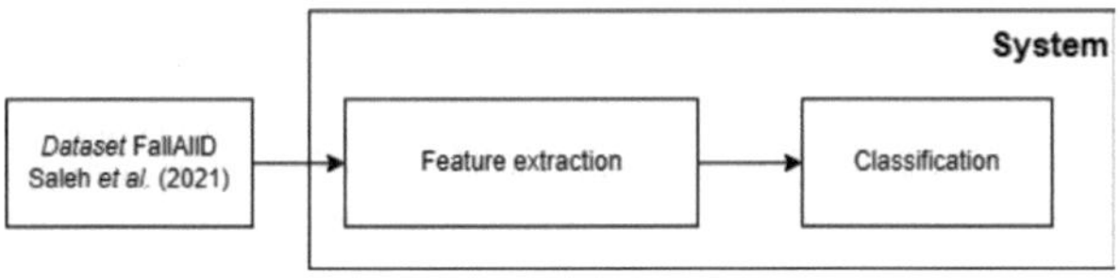

Figure 3.
Block diagram of the system.

due to its reliability and ease of use. Another functionality of this library is the ability to handle missing samples and to calculate simple statistical characteristics.

The Scikit-learn library was also used in this work. This library allows to create, train, and test the ML models. The Scikit-learn also release access to ML models, and to different training techniques, prediction, and allows to divide the dataset into training and test sets. The confusion matrices are also generated by a function of the Scikit-learn library. The Matplotlib was also used to plot the confusion matrices previously generated. Finally, Pickle allows developers to save and load datasets and ML models.

Datasets available to the academic community were researched. In Refs. [12–14] were found three datasets. The dataset in ref. [12] has the biggest data samples, however some miss relevant data. On the other hand, the dataset in ref. [13] does not have a pattern in the time domain of sensor readings. In this work, the dataset created in Ref. [14] was used. It was recently created and does not utilize mattresses to cushion the falls, making them more realistic. **Figure 3** depicts the block diagram of the proposed system.

The information extracted from the dataset contains the sensors readings from an accelerometer, a gyroscope, and a magnetometer. Next, the feature extraction was performed to train and validate the ML model. It is possible to perform the data classification after training the model, which can be done in two categories: Fall or Activity of Daily Living (ADL).

In Ref. [14], developers can capture data from the wrist, waist, and chest. The data captured from the waist was created by 14 volunteers, who used safety equipment to prevent injuries. The authors chose not to use mattresses to cushion the falls to be as realistic as possible. The volunteers were free to choose which ADLs or falls they desired to simulate. All 14 volunteers chose to simulate ADLs and 12 out of 14 volunteers performed simulated falls. Every single scenario was recorded for 20 seconds. During the first 9 seconds, the volunteer had a movement to simulate, when the ninth second was reached the volunteer mimicked a fall, and the person could stay down or recover depending on which type of fall he was simulating.

The authors labeled as ADLs or Falls data samples within the dataset by using numbers as activity IDs. IDs ranging from 1 through 44 are samples representing ADLs. Since we are only considering samples recorded by those sensors located in the waist, ADLs range from 13 through 44, because those activities labeled from 1 through 12 were recorded by sensors located in the volunteer's wrist. Among those ADLs, one can find activities such as: walking, running, standing up from a chair, and jumping.

Falls were labeled from 100 through 135. Among these falls, you can find different types of falls that normally would occur to people day to day. Volunteers simulated falls slipping, tripping, or losing balance while walking and slipping, and those falls were forwards, backwards, and laterally. They also simulated falls while running, lying in bed, trying to sit down, or standing for a while; these falls were simulated forwards, backwards, and laterally. It is also important to point out that falls with recovery were considered effectively as falls in this work.

It is important mention that those 14 simulating ADLs and those 12 simulating falls had to repeat the scenario several times to obtain the best and most accurate result. They could decide how much time they needed to rest between trials, and also, volunteers could decide the order in which they desired to perform the activities [5]. Repetition becomes a factor, as described in Ref. [5], because the volunteers can get used to the pattern of simulating that activity, resulting in activities performed in an unnatural manner.

With this said, we created a new column to label each sample as ADL or fall, represented by 0 s and 1 s, respectively. For this, we implemented a for loop, in which we compared the value stored in the activity ID column, and if this value was greater or equal to 100, we set the output column to 1, otherwise 0 was attributed.

Since the volunteers performed several times the same activity, the best scenarios were chosen to compose the dataset. Taking this into consideration, the dataset has 1797 samples of simulated falls and ADLs. Three features were extracted from the dataset to train the models: the mean, the standard deviation, and the range. The features were extracted for each one of the three axes of the sensors. The dataset was divided into three parts to perform training, validation, and testing of the models.

It is noteworthy to mention that the dataset needed simple data manipulation before extracting those features. The original dataset published by Ref. [14] is in bytes, so this way authors can adapt the dataset to their sensor's sensitivity. We considered the same sensitivity for the accelerometer, gyroscope, and magnetometer. The sensitivities were 0.244 mg/LSB, 70 mdps/LSB, and 0.14 mgauss/LSB, respectively. Since the dataset was used as a Pandas dataframe, we multiplied every column by its corresponding sensitivity; after multiplying every data sample, we obtained the sensor's original readings.

Figure 4 shows part of the Dataframe structure. It has 1798 rows and 7 columns in total. It is important to remark that only the data collected by the sensors located at the waist of the participants were used in this work. Also, the barometer readings were not considered.

```
FallAllD = pd.read_pickle("FallAllD.pkl")
FallAllD
```

✓ 2.4s

	SubjectID	Device	ActivityID	TrialNo	Acc	Gyr	Mag
0	1	Waist	13	1	[[4062, -1428, -45], [4190, -1422, -24], [4302...	[[1355, 21, 159], [1334, -19, 168], [1318, -68...	[[5541, 1551, 2887], [5568, 1587, 2956], [5605...
1	1	Waist	13	2	[[3737, -1549, 158], [3729, -1553, 160], [3728...	[[-4, 0, -1], [-7, 11, -3], [-11, 11, -2], [-1...	[[5185, 2842, 2267], [5220, 2876, 2260], [5211...
2	1	Waist	13	5	[[3954, -401, -107], [3894, -346, -108], [3835...	[[177, 173, 177], [160, 183, 164], [152, 197, ...	[[5626, 2148, 3057], [5602, 2127, 3015], [5562...
3	1	Waist	13	6	[[3992, -682, -33], [3988, -694, -31], [3985, ...	[[12, 13, 9], [15, 12, 10], [11, 12, 9], [9, 1...	[[5530, 2536, 2713], [5521, 2561, 2774], [5493...
4	1	Waist	13	9	[[2223, 167, -18], [2139, 222, -90], [2108, 25...	[[-396, -163, 33], [-338, -62, 54], [-272, 44,...	[[5752, 2071, 2723], [5801, 2220, 2615], [5776...
...	...	...	...	...	...	...	...
1793	15	Waist	131	2	[[3805, 827, 464], [3822, 855, 481], [3836, 80...	[[125, -157, -474], [114, -161, -460], [108, -...	[[2863, 617, 4182], [2891, 685, 4202], [2891, ...
1794	15	Waist	132	1	[[4019, -542, 564], [4014, -542, 564], [4016, ...	[[-19, 18, 12], [-21, 19, 10], [-24, 16, 11], ...	[[3567, 587, 1707], [3561, 493, 1684], [3551, ...
1795	15	Waist	134	2	[[3884, -1500, 557], [3875, -1485, 570], [3915...	[[42, 42, -8], [48, 37, -4], [49, 32, -1], [57...	[[2627, 2463, -442], [2725, 2403, -415], [2763...
1796	15	Waist	135	1	[[4045, -152, 405], [4043, -143, 398], [4040, ...	[[21, 7, 1], [22, 7, 3], [21, 9, 5], [22, 9, 3...	[[3521, 859, 3254], [3559, 857, 3262], [3553, ...
1797	15	Waist	135	2	[[4006, -691, 18], [4010, -708, 24], [4007, -7...	[[-2, 9, 6], [-5, 12, 4], [-4, 12, 5], [-8, 9,...	[[3319, 2445, 1284], [3349, 2405, 1268], [3325...

1798 rows × 7 columns

Figure 4.
Example of the Dataframe structure.

Acc	Gyr	Mag	Mean Acc X	Mean Acc Y	Mean Acc Z	...	Mean Mag X	Mean Mag Y	Mean Mag Z	Std Mag X	Std Mag Y	Std Mag Z	Rng Mag X	Rng Mag Y	Rng Mag Z	Fall
[[0.92842, 0.201788, 0.113216], [0.932568, 0.2...	[[8.75, -10.99, -33.18], [7.98, -11.2700000000...	[[0.40081999999999995, 0.08638, 0.58548], [0.4...	0.421164	0.486860	-0.259032	...	0.255893	0.063975	0.584274	0.155849	0.054347	0.116918	0.51660	0.49532	0.92008	1
[[0.980636, -0.132248, 0.137616], [0.979416, -...	[[-1.33, 1.2600000000000002, 0.840000000000000...	[[0.49937999999999994, 0.08217999999999999, 0...	0.477437	-0.463409	-0.259671	...	0.413253	0.238942	0.357015	0.084351	0.155681	0.124277	0.34916	0.38920	0.64218	1
[[0.947696, -0.366, 0.135908], [0.9455, -0.362...	[[2.9400000000000004, 2.9400000000000004, -0.5...	[[0.36777999999999994, 0.34481999999999996, -0...	0.516297	0.216274	-0.216580	...	0.337363	0.058966	0.370667	0.181615	0.105069	0.136420	0.43820	0.41636	0.78834	1
[[0.98698, -0.037087999999999996, 0.0988199999...	[[1.4700000000000002, 0.49000000000000005, 0.0...	[[0.49293999999999993, 0.12025999999999999, 0...	0.838070	0.347216	0.030255	...	0.470200	0.140514	0.245743	0.021936	0.126791	0.167907	0.13300	0.35028	0.64596	1
[[0.977464, -0.168604, 0.004392], [0.97844, -0...	[[-0.14, 0.6300000000000001, 0.420000000000000...	[[0.46465999999999996, 0.3423, 0.17975999999999...	0.197696	-0.241964	0.443883	...	0.173890	0.317003	0.045909	0.280613	0.035531	0.091014	0.61936	0.20300	0.38052	1

Figure 5.
The modified Dataframe structure.

The feature extraction was performed using the built-in functions in the Pandas' library. Pandas has a mean, standard deviation, and minimum and maximum functions available and ready to use, so firstly 27 new columns were added to the dataframe to save the features. Eighteen columns were needed since we are considering the three features for every one of the three axes, that is, three columns for acceleration mean in x, y, and z, repeating this to the standard deviation and range of the accelerometer, so having a total of 27 columns.

We transformed the original column of each sensor containing all three axes into three separated columns to represent each of them. Next, we used the functions mentioned before to calculate the features. Since there is no built-in function to calculate the range, we find the maximum value and subtracted the minimum value. After completing these steps, the dataframe has all the characteristics and it is ready to use with the ML model.

Figure 5 shows the Dataframe final state after including the accelerometer, gyroscope, and magnetometer features.

In **Figure 5**, it is possible to observe the pure data of the "Acc", "Gyr," and "Mag" sensors; however, the models will be trained with the columns that are on the right of those measures. A column called "Fall" identifies whether this event represents a fall or an ADL.

We used 80% of the data (not the 80% of the volunteers) for training, 10% of data for validation, and the residual 10% for final testing. It is important to note that the models were trained using the stratified k-fold cross-validation technique, with k

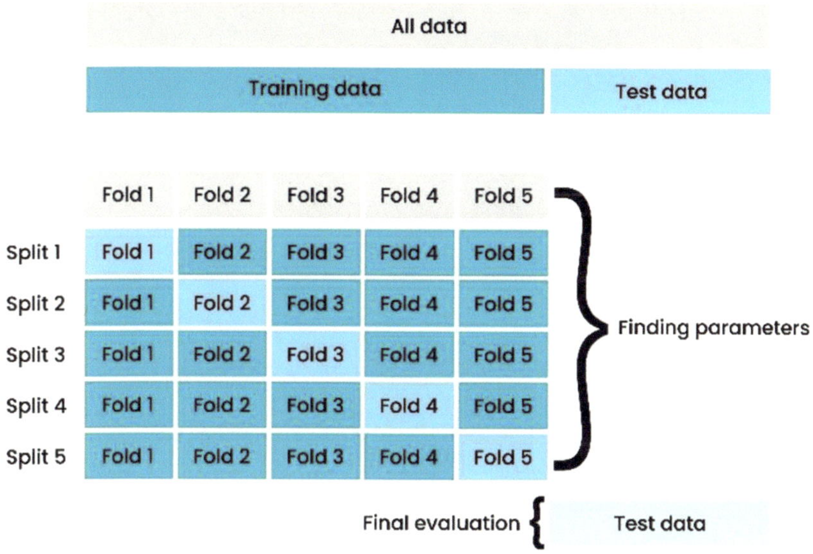

Figure 6.
k-fold cross-validation. Adapted from Ref. [17].

equal to 10, to obtain a good balance among the output classes. Data for both, training and validation, are divided into 10 parts, where k-1 is used for training and k is used for validation. This task is repeated k times to complete the training of the models. **Figure 6** illustrates this process.

To perform the data classification, two ML models were created. One model uses the SVM classifier, and the other one uses the k-NN. Both models use all the sensors' data with their respective characteristics. The models were studied and compared with the results obtained by the authors in the related works.

4. Results

4.1 SVM

In this work, the dataset was divided randomly. By performing the training and validation, it was possible to achieve an accuracy of 95.05% using the SVM model. However, this value cannot be considered the final accuracy because it is necessary to submit the model to a final test. In the final test, we used data which was not previously known by the model. The purpose of this procedure is to classify the unknown data.

The accuracy of the final test was 93.89%, with a sensitivity of 85.10% and a specificity of 96.99%. The accuracy informs how many samples were correctly classified. On the other hand, the sensitivity is the ability to predict the true positives of each available category and lastly, the specificity is the ability to detect the true negatives of each category.

The confusion matrix is shown in **Figure 7**. This matrix was created from the results retrieved from the final test. It is possible to observe the true negatives, false negatives, false positives, and true positives, where 0 represents ADLs and 1 represents falls.

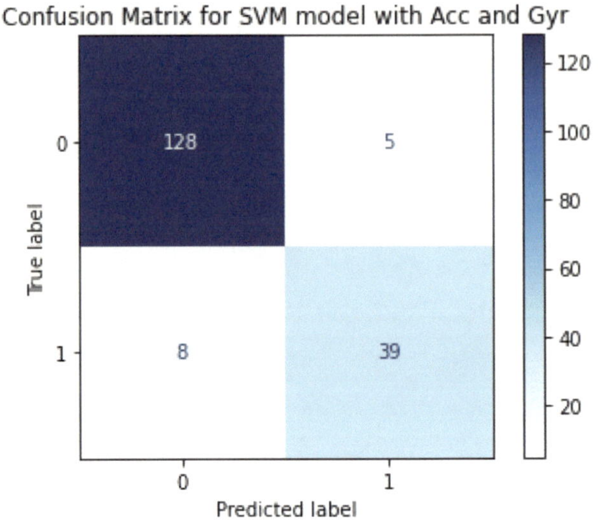

Figure 7.
Confusion matrix for the SVM model.

It is possible to observe that there were 129 true negatives, 4 false positives, 7 false negatives, and 40 true positives. This is a good result because the model correctly classifies 40 of the 47 falls.

4.2 *k*-NN

In this technique, we followed the same procedure as described before. After training, the model presented an accuracy of 88.45%. In the final test, with unknown data, it was possible to achieve an accuracy of 87.77%, a sensitivity of 82.98%, and a specificity of 89.47%. In this study, the results of the k-NN model were inferior, when compared with the SVM model, that is, the accuracy was 5.44% lower in relation to the SVM model. The confusion matrix for this model is shown in **Figure 8**.

Compared to the SVM model, the number of false negatives was increased by 1, and the number of false positives increased by 10.

4.3 Analysis

To get a better understanding of the results, it is necessary to make a comparison with the related works. It is worth noting that among the related works there is a discrepancy among the results using the different ML techniques. Likewise, it should be considered that each one of the authors used different features or methods to perform the data categorization (**Table 2**).

The best results can be found in Ref. [12] because the authors in Ref. [13] did not base their solution using ML. The classifier is based on thresholds; however, it is important to note that the extraction of the thresholds was performed using the SVM technique. In Ref. [12], the authors achieved accuracy of 99.1%, and in Ref. [14], the accuracy was 89.70% when applying the k-NN technique. In this chapter, we achieved an accuracy of 87.77%, thus 11.33% below the result of Ref. [12] and 1.93% below the results in Ref. [14]. The results obtained here are comparable to the results in Ref. [14] due to the similarity of accuracy between these studies.

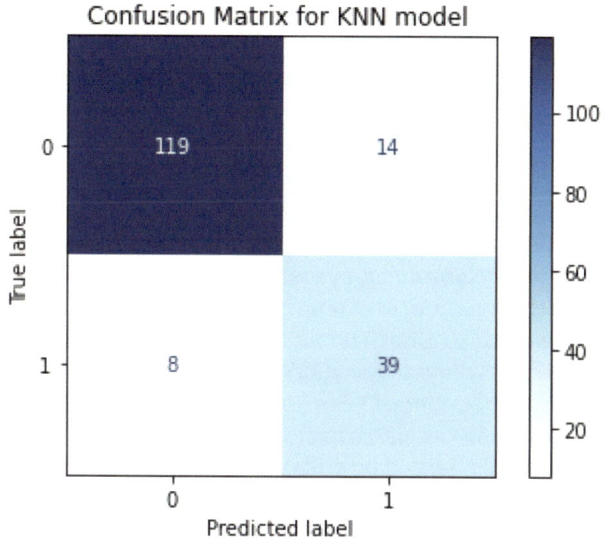

Figure 8.
Confusion matrix for the k-NN model.

Characteristics	[12]	[13]	[8]	[14]	This work
Sensitivity (%)	99.56	100	95.45	—	85.10
Specificity (%)	99.38	100	87.65	—	96.99
Accuracy (%)	99.48	100	93.25	84.66	93.89

Table 2.
Comparison between the best results achieved in the related works.

The SVM performed better in this work, so it is possible to make a direct comparison with Ref. [14]. In this chapter, the best accuracy was 93.89% compared with the 85.86% in Ref. [14]. In Ref. [12], the higher accuracy was achieved (99.48%). Different features were used for the accelerometer, magnetometer, and gyroscope sensors, considering that each one of the related works used different features to train the ML technique. In Ref. [14], the authors used three features, obtained from the accelerometer. In this work, we extracted three features, for each one of the three sensors.

Every work has its limitations, and this work is not an exception to that rule. The simple statistical characteristics can represent a limitation of this work. This can be considered as one due to its lack of precision representing the original signal. The original recorded signal was 20 seconds long, as mentioned before, so representing these signals only by using the chosen features can be not accurate enough. This limitation should be taken into consideration if the intent is having a more realistic classifier.

A shortcoming of this work is the fact that the ML models were not tested with a hardware implementation. All results were obtained by simulations, by using the test set separated in the first stage of the implementation. The models can be embedded into a microcontroller and classify data in real time. The outcome of a hardware implementation can yield different results, them being higher or lower in comparison with those obtained by simulation.

5. Conclusions

An ML-based approach to fall problem detection was presented in this work. A literature review made possible to understand what is behind a fall and its consequences on people, as well as to remark the ML techniques explored in the literature to approach this problem. In this study, two models were created using different ML techniques, and training was the same for both. We applied *k*-fold cross-validation, with training, validation, and testing sets. Both models were trained considering the data obtained from the accelerometer, gyroscope, and magnetometer.

The mean, standard deviation, and range were used as input features for the ML models. The results reached a value that enables comparisons to those in the related studies. The best result was accuracy of 93.89% for the SVM technique. Currently, an embedded system is being developed with an ESP32 microcontroller to communicate with the sensors, embedding the classification algorithm and sending notifications.

This work can be complemented by embedding the ML models and building a physical device to test the models in real time with sensor readings, consequently obtaining more realistic results. To further improve this work, we recommend employing more features, like authors of Ref. [12] did with their work. By applying more characteristics it is possible to have better results, since there is more information regarding the sensor's readings. Having more information fed to the models is a better approach because they can have a better understanding of what those characteristics are representing; therefore, a better division of possible outputs is achieved.

Acknowledgements

We gratefully acknowledge the support of the Fundação de Amparo à Pesquisa do Estado de Santa Catarina (FAPESC), grant number 2021TR001236; and the National Council for Scientific and Technological Development (CNPq), grant numbers 305835/2021-1, 424937/2021-2.

Author details

Teddy Ordoñez Nuñez, Raimundo Celeste Ghizoni Teive
and Alejandro Rafael Garcia Ramirez*
Universidade do Vale do Itajaí, Itajaí, Brazil

*Address all correspondence to: ramirez@univali.br

References

[1] World Health Organization. Falls [Internet]. 2021. Available from: https://www.who.int/news-room/fact-sheets/detail/falls. [Accessed: May 20, 2020]

[2] Cabral DraK de N. Quedas dos idosos podem ser prevenidas [Internet]. 7 de fevereiro. 2019. Available from: https://www.hospitalsiriolibanes.org.br/sua-saude/Paginas/prevencao-quedas-idosos.aspx. [Accessed: November 13, 2020]

[3] Lord CJ, Colvin DP. Falls in the elderly: Detection and assessment. In: Proceedings of the Annual International Conference of the IEEE Engineering in Medicine and Biology Society. Vol. 13. New York, USA: IEEE; 1991. pp. 1938-1939

[4] Williams G, Doughty K, Cameron K, Bradley DA. A smart fall and activity monitor for telecare applications. In: Proceedings of the 20th Annual International Conference of the IEEE Engineering in Medicine and Biology Society. Vol. 20. Biomedical Engineering Towards the Year 2000 and Beyond (Cat No98CH36286). New York, USA: IEEE; 2000. pp. 1151-1154

[5] Noury N, Fleury A, Rumeau P, Bourke AK, Laighin GÓ, Rialle V, et al. Fall detection - principles and methods. In: Annual International Conference of the IEEE Engineering in Medicine and Biology - Proceedings. New York, USA: IEEE; 2007. pp. 1663-1666

[6] Leite GV. Detecção de Quedas de Pessoas em Vídeos Utilizando Redes Neurais Convolucionais com Múltiplos Canais [Internet]. Universidade Estadual de Campinas. 2020. Available from: http://repositorio.unicamp.br/bitstream/REPOSIP/341843/1/Leite_GuilhermeVieira_M.pdf

[7] Rodrigues Silva EK. Desenvolvimento de um sistema de detecção de quedas para idosos [Internet] [Dissertação]. Universidade Estadual da Paraíba. 2018. Available from: http://tede.bc.uepb.edu.br/jspui/handle/tede/3570

[8] Júnior CSDQI. Sistema de Detecção de Quedas de Idosos [TCC]. RS, Brazil: Universidade de Caxias do Sul; 2016

[9] Abbate S, Avvenuti M, Corsini P, Light J, Vecchio A. Monitoring of human movements for fall detection and activities recognition in elderly care using wireless sensor network: A survey. In: Wireless Sensor Networks: Application-Centric Design. London: InTech; 2010. pp. 147-166. Available from: http://www.intechopen.com/books/wireless-sensor-networks-application-centric-design/monitoring-of-human-movements-for-fall-detection-and-activities-recognition-in-elderly-care-using-wi

[10] National Health Service. Overview: Falls [Internet]. 2021. Available from: https://www.nhs.uk/conditions/falls/. [Accessed: May 15, 2021]

[11] Mobilize. Relatório final da campanha e estudo realizado pelo Mobilize Brasil [Internet]. 2013. Available from: https://www.mobilize.org.br/midias/pesquisas/relatorio-calcadas-do-brasil---jan-2013.pdf. [Accessed: May 18, 2021]

[12] Özdemir AT, Barshan B. Detecting falls with wearable sensors using machine learning techniques. Sensors (Switzerland). 2014;**14**(6):10691-10708

[13] Pierleoni P, Belli A, Palma L, Pellegrini M, Pernini L, Valenti S. A high reliability wearable device for elderly

fall detection. IEEE Sensors Journal. 2015;**15**(8):4544-4553

[14] Saleh M, Abbas M, le Jeannes RB. FallAllD: An open dataset of human falls and activities of daily living for classical and deep learning applications. IEEE Sensors Journal. 2021;**21**(2):1849-1858

[15] Albert M. V.; Kording, Konrad; Herrmann, Megan; Jayaraman, Arun. Fall classification by machine learning using Mobile phones. PLoS One. 2012;**7**(5):1-6. DOI: 10.1371/journal.pone.0036556

[16] Rastogi S, Singh J. A systematic review on machine learning for fall detection system. Computational Intelligence. 2021;**37**(2):991-1014. DOI: 10.1111/coin.12441

[17] Scikit-learn. 3.1. Cross-validation: evaluating estimator performance [Internet]. 2022. Available from: https://scikit-learn.org/stable/modules/cross_validation.html. [Accessed: August 16, 2022]

Chapter 6

Machine Learning and Cognitive Robotics: Opportunities and Challenges

Thomas Tawiah

Abstract

The chapter reviews recent developments in cognitive robotics, challenges and opportunities brought by new developments in machine learning (ML) and information communication technology (ICT), with a view to simulating research. To draw insights into the current trends and challenges, a review of algorithms and systems is undertaken. Furthermore, a case study involving human activity recognition, as well as face and emotion recognition, is also presented. Open research questions and future trends are then presented.

Keywords: neural networks, cognitive control architectures, software frameworks, imitation learning, reinforcement learning

1. Introduction

Cognitive robotics aim at endowing robots with intelligent behaviour by providing processing architecture that allows them to interact with the environment, learn, understand and reason about the environment, and behave like humans in response to complex world dynamics. These are problem-solving, intentional (planning), reactive, learning, understanding and explaining behaviours. Behaviours are based on modelling biological systems, optimal control theory (engineering), neurosciences, and other behavioural sciences. Typical applications where cognitive capabilities are important in manufacturing are pick and placement, machine inspection, and collaboration and assistance. Service robots are specialized robots [1], which operate either semi or fully automatically to perform services useful to humans (excluding manufacturing operations), such as caring for the elderly and rehabilitation. The autonomy of such robots is fully oriented towards navigation in human environments and/or human-robot interactions. Enabling more autonomous object manipulation with some level of eye-hand coordination and high precision in a complex environment is a challenge [2]. To embed systems with more sense of intelligence, collaborations between AI, machine learning and robotics communities are essential to achieve remarkable progress. Robot learning refers to the robot learning about itself and the effect of its motor commands and action. Examples include learning sensorimotor skills (locomotion, grasping and object manipulation) or interactive skills (manipulation of an object in collaboration with a human being). The field of developmental robotics and evolutionary robotics has also emerged to deal with how robots learn. In cognitive robotics, an integrated view is taken of the robots, their motor, perceptual subsystems and the body's interaction with the environment. The main challenge is a lack of adequate knowledge of the human brain at different stages of development to enable adequate modelling.

Mobile agents are the principal means of embedding cognitive processing capabilities in robotic systems. These are software components that can carry out functions autonomously on behalf of another entity to realize tasks and can migrate from one robot to another through Wi-Fi networks. Embedded cognitive robotics focuses on understanding and modelling perception, cognition and action in artificial agents through bodily interactions with the environment to be able to perform cognitive tasks autonomously [3]. Several authors have reported works using mobile agents [4, 5]. From a technical point of view, there are several open challenges in the implementation of motor and cognitive skills in artificial agents. State-of-the-art robots are still not properly able to learn, adapt, react to unexpected conditions and exhibit a level of intelligence to operate in an unconstrained environment.

Machine learning (ML) algorithms are computationally intensive data-driven analysis, modelling and inference techniques based on statistical (clustering), evolutionary computing, neural networks (deep neural networks) and mathematical optimization [6]. The processing pipeline given a set of data sequentially consists of preprocessing, feature extraction, modelling, inference and prediction. The modelling stage may involve iterative minimization of the criterion of the model fit between a discriminant and the data. It focuses on the development of algorithms that allow computers to automatically discover patterns in the data and improve with experience, without being given a set of explicit instructions. ML has been applied in experimental robotics to acquire new skills; however, the need for carefully gathered data, clever initialization and conditioning limits the autonomy with which behaviours can be learned. In particular, deep learning neural networks with several levels of composition have achieved remarkable performance in vision and natural language processing. It can be leveraged via transfer learning to generalize from simulation to the real world via domain randomization [7–9] to learn end-to-end visuomotor controllers [10, 11]. The limitations of deep neural network (DNN) techniques such as interpretability, susceptibility to adversarial attacks, privacy issues and stability under perturbations in designing end-to-end control policies are worth addressing. In particular, reliable long-term prediction is desirable to enable re-planning to adapt to the changing environment [12].

Machine learning techniques embedded within current AI systems (via agents) have increasingly shown sophisticated cognitive capabilities. For example, an existing approach in machine learning to lexicon acquisition is focused on symbol grounding problems on how to connect sound information from a human and sensor information from robots captured from the environment. A multi-sensory approach based on co-occurrence probabilities between words and visual features that is observed by a robot [13] improves as a result of using an active selection of motion based on saliency [14, 15]. Several developments in cognitive robotics underlying its multi-disciplinary nature are presented.

Traditional approaches of processing are based on a bottom-up approach with the processing pipeline starting sequentially from sensing, perception, cognition and action under control architecture such as in ref. [16], which is essentially behaviour based and, later on with high-level decision processing [17], incorporated to enable more autonomy. Fundamental to robotics are the control policies that guide the behaviour of a robot. It is mainly based on control theory and mathematical optimization or biologically inspired models with control relying on vision in combination with other sensing modalities (olfactory) [18, 19]. A lot of models have been developed governing the behaviour of a robot itself and how it interacts with its environment [12, 20, 21].

There are three main control architectures, namely, logic-based, subsumption and hybrid architectures [16]. The logic-based architecture uses a set of rules and provides pro-active behaviour, whilst the latter incorporates intelligence and interaction with the environment as a means of introducing cognition. Behaviour is organized hierarchically. The hybrid architecture achieves modularity and interactivity between layers. Because the models used were relatively simple, it suffers from the problem of scalability and modelling of complex scenarios. Instead of providing all information to the robot a priori, for example, possible motions to reach a certain target position, the agent will, through some process, 'learn' which motor commands lead to what action. For autonomous systems, a decision level incorporating capacities of producing plans and supervising their execution, whilst at the same time being reactive to events from the previous layer has been added to the top-level hierarchy [6]. They are typically used in controlling the robot (motion control) or in carrying out tasks. Different multi-robot configurations including robotic swarm use multi-agent systems to carry out complex tasks.

Predictive processing (PP) [3], a processing approach in cognitive sciences, is increasingly being used in cognitive robotics. It is a top-down approach that aims at unifying perception, cognition and action as a single inference processing. It is predominantly based on the free-energy principle [22], which is associated with frameworks such as predictive coding, active inference and perceptual inference. The free-energy principle seeks to minimize prediction errors [20]. It asserts that through bodily interaction with the environment, agents are expected to learn and then be capable of performing cognitive tasks autonomously [23]. The core of information flow is top-down and the bottom-up flow of prediction error. Control motor commands are replaced by proprioceptive top down prediction using the forward model [24]. PP is typically used in motor control and estimation of body states of a robot [25, 26]. A neural network is typically used as the generative model. Active inference, a related frame work, aims at minimizing prediction error or free energy using variational inference. It involves constructing a forward model involving hidden states to reduce proprioceptive noise for control [21].

To address issues in cognitive robotics, researchers in developmental robotics build artificial systems capable of acquiring motor and cognitive capabilities by interacting with the environment inspired by human development [27]. Traditionally mobile agents, simulated robots, humanoids or specially designed apparatus are used for research into higher-order cognitive capabilities (learning, communication and understanding) mimicking the functionalities of the human brain like its internal structure, infrastructure and social structures. The model starts from foetal sensorimotor mapping (mechanisms of dynamic motions and motor skill development) in the womb, body and motor representation and spatial perception through to social behaviour learning (communication, action execution and understanding) and spatial perception. Important insights have been gained; for example, ref. [28] indicates that control and body structure are strongly connected, with the body having the role of controlling its motion. In ref. [29], dynamic walkers realize walking on slopes without any explicit control or actuation, saving energy.

Models of human communication mechanisms have been used in developing interactions such as between caregivers and robots, action execution and understanding, development of vocal imitation and joint attention [27] in human-robot communications. Sumioku et al. [30] proposed an open-ended learning loop of social action by which artificial infants reproduce experience contingency using information-theoretic measure of contingency. Typically, gaze-following or utterances about the focus of attention are used for joint attention. Human-like robots able to show distinct facial expressions to be used in specific situations have been developed [31, 32], but the robots are unable to adapt to non–pre-specified situations. From control perspective, some of the capabilities required [33] for collaboration and assistance between robots and humans are as follows: the ability to perceive the world in a similar way to humans; the ability to communicate with humans using natural language; the ability to develop cognition through sensorimotor association; the ability to use attention and emotion to control behaviours and the ability to produce appropriate behaviours in a variety of situations. Clearly, this calls for a multidisciplinary approach involving neural sciences, developmental robotics, psychology, and engineering. Kawamura and Brown [34] approached the problem using working memory-based multi-agent systems for robot behaviour generation.

Evolution has equipped humans with a wide range of tools for collaboration, including the use of language, gestures, touch, and facial expressions, to facilitate interactions. Robots must support many of these communication methods to effectively collaborate or assist humans. In particular, for robots working in human environment, there is an urgent need to anticipate and recognize bodily movements and facial expressions, to offer timely and effective assistance when needed. To this end, a case study involving facial and action recognition to illustrate some capabilities in this regard is presented. The rest of the chapter is structured as follows: Section 1.1 introduces computational architecture and platforms for cognitive robotic systems. Sections 1.2 and 1.3 cover the roles of technology and software, respectively. Section 1.4 deals with the role of decision-making in cognitive systems. Sections 1.5, 1.5.1 and 1.5.2 briefly introduce the main algorithms used in cognitive robotics, namely, reinforcement learning and imitation learning algorithms, highlighting developments in ML that have made it possible for renewed interest in these algorithms. Section 1.5.3 reviews deep learning networks for feature learning and classification. Sections 1.6 and 1.6.1 provide a case study on human activity recognition. Section 1.7 briefly reviews current trends, whilst Section 1.8 discusses successes, challenges and research directions. Finally, Section 1.9 concludes the chapter.

1.1 Architecture and platforms for cognitive robot research

To facilitate the development of mature cognitive robotic systems, several computing platforms including real robots like humanoids (icub), panda and Hobo, simulators, and middleware like ROS and YARP are available. Particularly, to facilitate the development of mature cognitive systems, robots must continuously interact with the environment, know where objects are in the scene and understand the consequences of their generated actions. The icub [35] humanoid robot is a 53-degree-of-freedom humanoid robot of approximately the same size as a three-year-old child. It can crawl on all four limbs and sit up. Its hand allows dexterous manipulation, and its head and eyes are fully articulated. It is an open systems platform available for research under GNU general public license. Its capabilities are built based on an ontogenetic pathway of human development. **Figure 1** shows different postures of icub. Robotic simulators are of interest despite not being able to provide a full model

Figure 1.
iCub robot in different postures from ref. [35].

of the complexity present in the real environment. For example, the icub simulator [35] has been designed to reproduce as accurately as possible the physics and dynamics of the robot and its environment with the constraint of running approximately in real-time. It is composed of multiple rigid bodies connected via joint structures. It consists of the following components: physics and rendering engines, YARP protocol for simulated icub and body model. All commands sent to and from the robot are based on YARP instructions. More details are provided in ref. [36]. Besides, there are several platforms for humanoids and other robots in studies reported in ref. [37–39]. Details of Pioner3-AT bender robotic platform are provided in ref. [40]. There are also several European Union funded research projects on cognitive robotics that have resulted in several architectures, system concepts and benchmark datasets [41]. Several simulators for robotic systems are provided in ref. [42]. To build cognitive systems, several computational architectures have been designed and built to realise different cognitive platforms.

The following are representative architectures: The Clarion [43–45] architecture is a broadly scoped computational psychological model based on the dual theory of the mind, capturing essential structures mechanisms and processes of the mind. It provides a framework, essential structures and computational model for realising processes of the mind. It also facilitates detailed exploration of the mind and psychological theories. Clarion consist of four subsystems, namely, action-centred subsystem (ACS), non–action-centred subsystem (NACS), motivational subsystem (MS) and metacognition subsystem (MC). MS provides the impetus for action and cognition, whilst MC provides for monitoring and regulating other processes. Together, these subsystems address action, skill learning, memory, reasoning, motivation, personality, emotions and their interactions. **Figure 2** is a high-level diagram of Clarion. Each subsystem consists of two levels, which is a dual representation structure. The top-level encodes

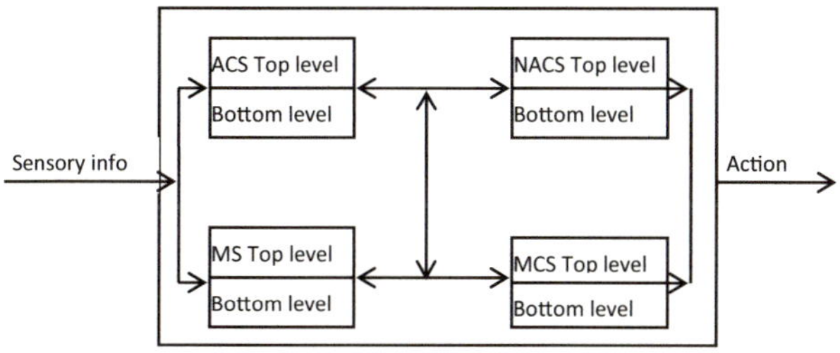

Figure 2.
CLARION architecture [44].

explicit knowledge, potentially corresponding to 'conscious', and the bottom layer encodes implicit knowledge corresponding to 'unconscious' knowledge and also corresponds to symbolic versus connectionist representation.

Computationally, ACS is realised with multilayer perceptron or reinforcement learning, whilst NACS implements implicit declarative processes with associative memory. Explicit declarative processes are captured as symbolic associative rules. Implicit processes deal with drive activations captured by MLP and explicit processes deal with goals. More details of the architecture are provided in ref. [44]. Other general purpose architectures include Soar [46], which integrate knowledge, intensive reasoning, reactive execution, hierarchical reasoning and learning from experience. It has the goal of creating systems with cognitive capabilities like humans. Several other projects that target specific robotic platforms have produced application-specific cognitive architectures. These include the HAMMER [47, 48] and ArmarX and Xperience architectures on Armar humanoid robot [49]. The HAMMER architecture is for assistive robotic agents cooperating with humans to carry out tasks. It provides for sensing user states and actions, modelling skills and predicting intentions and personalising to maximise assistance effectiveness over extended periods of interactions. ArmarX is a hybrid architecture, proposed for human observation and experience. Interactions with humans occur in natural language. It recognises the need for help and reason about the world. The original architecture proposed has been continuously extended in several projects. It consists of three layers, namely, high-level layer for planning and reasoning, mid-level layer for mediating symbolic knowledge and sensory-motor data; and low-level layer for robotic behaviour focusing on functions and skills, hardware abstraction layer and bridging middleware to other robot software frameworks.

Using virtual environments for simulation is very important to ensure the safety of robots, humans and other objects in the environment; the slow wall clock time makes it a too slow method to generate enough data in a reasonable time frame, and physical trials are slow and costly and the learned behaviours are limited. Increasingly, the use of complex simulation environments is being used for experimentation and research. By training a virtual robot in countless situations, such as low-probability scenarios, it is the objective of the system to learn to generalize from the scenarios and safely handle future yet unseen scenarios. When the physical properties of the environment, such as gravity, friction coefficients and the object's visual appearance, are used and randomized, it becomes apparent that the learned models transfer successfully to the physical robot using domain randomization [50]. One such platform is the Unity [51, 52] 3-D rendering platform, a cloud scalable infrastructure for generating thousands of frames per second. For video games, Arcade learning environment (ALE) [53] is a standard test bed for deep reinforcement learning (DRL) algorithms, and it supports discrete actions. TORCS car racing simulator [54], on the other hand, supports continuous actions for deep reinforcement algorithms.

1.2 The role of technology

The pervasiveness of information communication technology (ICT) is evident everywhere in our daily lives. In industrial settings, the following are some examples:

- Monitoring and control of all tools of production;
- Collecting data for many sensors for monitoring, control and predictive maintenance of equipment;

- Use of machine intelligence, wireless connectivity, cloud computing to integrate physical systems and processes (Industry 4.0 effort);
- Use of key enabling networking technologies, namely, edge, fog and 5G using AI agents;
- Applications of service robots (non-industrial) include shopping, travelling, home assistance and elderly care.

In our daily lives, examples include the numerous gadgets in our homes to assist in our daily lives and care for the elderly. Robotics network and cloud robotics have evolved to connect robots and allow a central or distributed intelligence to command and control any set of robots. Advantages include flexibility, simplification of hardware and software about the robot, ease of re-planning and task management of complex robots. Several configurations exist for robots, namely, stand-alone robots, networked robots and cloud robotics [55]. Networked robots address the problem associated with stand-alone robotic systems by sharing perceived data with each other and executing tasks in a cooperative and coordinated manner. Cloud computing empowers robots by providing faster and more powerful computational capabilities through massively parallel computation (using CPUs, GPUs, and clusters and data centres) and higher storage facilities, as well as access to open source, big datasets and software cooperative learning capabilities.

Typical applications include human-assisted driving and self-driving vehicles for safe transportation, Industrial 4.0 drives to create cyber-physical systems for industrial processes based on cloud by creating a replica in the cyberspace for closed-loop feedback [56] and support for autonomous and smarter processes. It also caters for the convergence of sensing, computation and communication by providing a common platform for integrating data acquisition, processing, storage and decision making. AI agents for digital twin 4.0 provide movement prediction, tasking learning, risk reduction and predictive maintenance. Fundamental to most of these developments is AI and ML for continuous decision-making.

CR are expected to continuously learn and adapt to their environments and make decisions in real-time when required under conditions of uncertainties in sensor data, processing complexities, privacy and security constraint to arrive at timely and effective decision-making. AI and ML empowered agents is one approach to realising this goal. Current robotics have made significant progress in sensing perception and control problems but find it challenging to provide integrated thinking, feeling and knowing [57]. It is still very challenging for two-legged robots to walk naturally in unconstrained environments. Several challenges exist in using robotic platforms such as the high cost of prototyping, steep learning curve and programming robots to carry out complex tasks like autonomous driving in unconstrained dynamic environments.

1.3 The role of software

Closely related to cognitive robotics is cognitive computing (CC), which is a multidisciplinary field aiming at devising computational models and decision-making mechanisms based on neurobiological processes of the brain, cognitive sciences and psychology. It aims to endow computers with the ability to think, feel and know. Since there is no commonly accepted definition of cognition, there are several definitions of cognitive computing [58, 59]. Wang [59] defines cognitive computing in terms of cognitive informatics that applies how the brain processes information and copes with decision-making to information sciences. CC is defined as an emerging paradigm of intelligent computing methodologies and systems based on cognitive informatics that implements computational intelligence by autonomous inferences and perceptions mimicking the mechanisms of the brain. Research in cognitive computing is focused on three thematic areas, namely, computer systems with a faculty of knowing, thinking and feeling. Applications of CC include education, healthcare, commerce and industry.

When software adds intelligence to information-intensive processes, it is known as robotic process automation. The process uses AI to extend and improve action and saves cost and customer satisfaction. It is typically used in completing a complex business process that uses unstructured data or persists over a long period [57]. Typically a bot (an agent for a user of a program) observes the process to automate the process.

One of the requirements for robust and effective CR is software integration frameworks. This is justified when one considers the following:

a. Cognitive models are derived from a large spectrum of computational paradigms that are not necessarily compatible when considering the underlying software architecture;

b. Changes in application requirements due to hardware interfaces, computational and network latencies and the need for integration;

c. Cognitive research projects utilize robotic systems as demonstrators, and therefore serve as an important proof of concepts and might also require integration;

d. The need to provide common interfaces and functions;

e. Specific software frameworks may be required to take advantage of innovations in hardware (new development of brain-like hardware architecture) and the development of relationships among concepts of a given domain.

Software frameworks enable thinking by taking advantage of brain-like computer machinery or determine causal relationships among concepts of a given domain. There have been several published works on software frameworks [60] prototyping, development of middleware, sustainable software design and architectural paradigms. MARIE [61] is a component-based software architecture for integrating and combing heterogeneous software and computational paradigms. It adapts the mediator design pattern to create a mediator interoperability layer (MIL). MIL is implemented as a virtual space where applications can interact together using a common language. ROS (robot operating system) [62], an open-source robotic middleware suite, is frequently used in robotic projects. ROS provides a set of software frameworks for software development. ROS provides the following services: hardware abstraction, low-level device control, message passing between two processing, package management and other functions; ROS 2 [63] and above provide real-time support and an embedded system. ROS is made up of three components: language and platform independent tools for building and distributing ROS-based systems; ROS client implementations (Roscpp, rospy, roslisp, etc) and packages containing application-related code. Ros typically connects to robots via webSockets and operates on cloud servers. There are several platforms on which ROS runs including ROSbot, Nao Humanoid [64] and Raven II surgical robotic research platforms. Peira et al. [65] provide a framework for using ROS on the cloud. Davinci [66] is another software framework that is cloud-based for service robots exploiting parallelism and scalability. It is based on the Hadoop cluster combined with ROS as the messaging framework. Fast SLAM algorithm, an environmental mapping algorithm for large-scale mapping, was implemented on this platform with significant performance improvement.

A framework for unifying multi-level computing platforms and orchestrating heterogeneous edge, fog and cloud computing resources compliant with MEC [67] was proposed in ref. [68]. It is suitable for integrating different computing, communication and software technologies.

1.4 The role of decision-making

At the core of most ML tasks is decision-making based on information fed to the decision maker, for example steering or breaking a car. Decision-makers used to be either a human or a group of humans; now it can be AI using different combinations of ML and traditional algorithms via agents technology. According to Kahneman [69], there are two modes, namely, system 1 and 2 modes of the human brain, and most ML methods emulate the mode of operation in system 1. ML establishes empirical associations through training and learning. When given scenarios resembling training scenarios, ML yields results in a fast way. However, it struggles when given scenarios not covered during training or the training was inadequate. In human decision-making, when system 2 fails to intervene because it is fooled by an apparent coherent picture created by system 1 tends to result in decision-making. Thus, if ML is to be used in decision-making, the ability to detect difficult and dangerous situation tend to trigger system 2.

In cardiovascular medicine, ML is routinely used to perceive an individual by collecting and interpreting his/her clinical data, and clinicians would reason on them to suggest actions to maintain or improve the individual's health. Thus, it mimics the clinicians' approach when examining and treating sick patients [70]. Big data leveraged by ML can provide well-curated information to clinicians so that they can make better informed diagnosis and treatment. ML analyses have demonstrated human-like performance in low-level tasks in robotics and cardiology.

There have been studies reporting on the success of sensing-perception-control/action loop in autonomous vehicles [56].

Higher-level tasks involving reasoning such as patient status interpretation and decision support, and reasoning under uncertainties and dynamic environment in robotics have proven to be challenging. Intention predictions in a dynamic environment are also challenging. Similarly, human-robot cooperation for safe road transportation includes challenges in infrastructure [71] (sensor, communication subsystem, computing and storage) and predicting behaviour when driving, motion prediction and gesture recognition.

1.5 Review of algorithms

From the cognitive architecture descriptions discussed, at the high level, the actions of a robot are goal-directed, with the middle layer responsible for intermediate organisation, planning and execution using some memory hierarchy. The bottom layer is reactive and deals with the environment. For a robot to be able to interact with other objects and its environment, it needs to know how to predict the consequences of its actions using typically a forward model: X=(S, π_θ), where s is the state of the robot, and π_θ: S->A is a parameterized action policy (A) to the space of effect or task space. Similarly, the inverse model computes the action policies that can generate a given effect (S, Y)-> π_θ. Some examples are mapping of movements of the hand in the visual field to the movement of the end point of a tool, and oscillation of the legs to body translation of a robot. There are two main approaches, analytical approach based on control engineering and learning-based approach. The main challenge is to model a prior all the possible interactions between a robot and its environment. Learning is additionally confronted with multimodal sensing perception, high dimensional spaces, continuous and highly non-stationary spatially, and temporary state spaces. Typically, statistical regression is used to guide autonomous exploration and data collection. Alternatively, an approach for learning and constraining the environment is active learning. Several learning paradigms have been used including reinforcement learning and imitation learning. Several machine learning techniques such as deep learning have been used to model robotic agents in the real world. Deep learning networks build a model that produces end-to-end learning and inference system driven purely by data. Most of the approaches reported in the literature make use of neural networks to construct forward and inverse modules. To overcome the problem of catastrophic forgetting (training a model with new information interferes with previously learned knowledge [72]) in neural networks, special memory architectures may be used [34] besides pure algorithmic approaches. Additionally, other cognitive approaches from developmental robotics, neuroscience and other behavioural science approaches have been used. Active learning and inference approaches constrain the search space and allows self-exploration. These methods generally begin using random and sparse exploration, build meta-models of the performances of the motor learning mechanism and concurrently guide the exploration of various subspaces for which the notion of interest is defined [73]. Interest is defined in terms of variants of information gain (variance, entropy or uncertainty). Motivational and goal-driven approaches where exploration and search are goals/curiosity or attention driven [74–76] to reduce the large search spaces. Cognitive processing techniques can be split into two main approaches, namely, the control theory approach and the free energy-based approach. Although both of them use optimization techniques, the latter approach seeks to minimize free energy prediction error using variational or Bayesian approaches.

1.5.1 Review of reinforcement learning

There are three main classes of algorithms for machine learning, namely, supervised, unsupervised and reinforcement learning. In supervised learning, data defining the input and corresponding output (often called 'labelled' data) are available. In unsupervised learning, only the input is available and the structure of the underlying data is typically solicited. It is used to explore the hidden structure of the data. In reinforcement learning (RL), learning takes place by trial and error interactions with the environments. It is goal-directed learning that constructs a learning model specifying output to maximize long-term profit. Deep RL (DRL) uses deep learning methods (multi-layer neural network) to learn models and representations at different levels of abstraction [77] in an unsupervised manner. It leverages deep learning as a function approximator to deal with high-dimensional data. DRL algorithms have been applied to robotics allowing control policies for robots to be learned directly from camera inputs in the real world [11]. The basic model of RL is shown in **Figure 3**.

At time t, the agent receives state s_t from the environment. The agent uses its policy to choose an action a_t. Once the action is executed, the environment transitions a step providing the next state S_{t+1}, as well as feedback in the form of reward R_{r+1}. The agent uses knowledge of state transitions of the form (S_t, A_t, S_{t+1}, R_{t+1}) to learn to improve its policy. A policy (π) is a mapping function from any perceived state s to action taken from that state. Alternatively, a policy can be interpreted as a probability distribution of candidate actions that will be selected from state (s) as in Eq. (1):

$$\pi = \phi(s) = \left\{p(a_i|s)|\forall a_i \in \Delta_\pi \Lambda \sum p(a_i|s) = 1\right\} \quad (1)$$

Δ_π denotes candidate actions on policy π, and $p(a_i|s)$ denotes the probability of taking action a_i given the state s. A policy is deterministic if the probability of choosing an action a from s is p(a/s)=1 for all state s, otherwise, stochastic, i.e, p(a|s)<1. A value function is used to evaluate how good a certain state or state-action pair (s,a) is. For this purpose, a generalized return value R_t, defined by Eq. (2) is used, where γ (0 < γ<1) is the discounted factor.

$$R_t = r_{t+1} + \gamma r_{t+2} + \gamma r_{t+3} \ldots \gamma^{T-t-1} r_t = \sum_{i=0}^{T-t-1} \gamma^i r_{t+i+1} \quad (2)$$

The value of a state under policy π is evaluated as the expectation of R_t defined by Eqs. (3) and (4) for the state and state-action pair, respectively. E denotes expectation operation.

$$V_\pi = E[R_r|s_t = s, \pi] \quad (3)$$

$$Q_s(s, a) = E[R_r|s_t = s, a_t = a] \quad (4)$$

Underlying RL is dynamic programming [78] and bellman equations for optimality under Markov decision process modelling. RL algorithms have been successfully applied to several real-world problems with limited state spaces to problems in control and navigation. However, it faces the following challenges:

- The optimal policy must be inferred by trial and error interaction with the environment with the only learning signal being the reward.

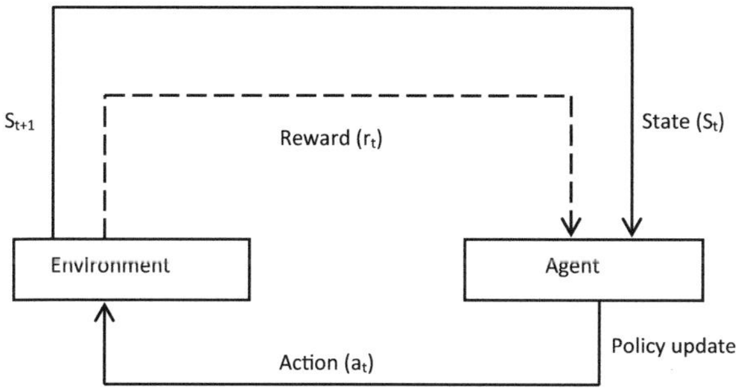

Figure 3.
RL algorithm using a single agent.

- Since the observations of the agent depend on its action, it may contain strong temporal dependencies.
- Long-range dependencies may only emerge after many transitions.
- Balancing exploitation versus exploration.

Underlying RL is the Markov property that the current state affects the next state or is conditionally independent of the past given the present state. Partially observable Markov decision processes (POMDP) are Markov decision processes (MDP) in which agent receives an observation $p(o_{t+1}|s_{t+1},a_t)$ where the distribution is dependent on the current state and previous action [79]. An episodic MDP resets after each episode of length T, and the sequence of states, actions and rewards in an episode constitute a trajectory or rollout of the policy. There are three main types of reinforcement algorithms, namely policy-search, value-function based and those that combine both policy and value function approaches. They include actor-critic method, temporal difference and Monte Carlo-based methods [80, 81]. The increasing use of deep reinforcement learning (DRL) algorithms has been attributed to the low-dimensional representation of deep neural network representation and the powerful functional approximation of neural networks. The following significant recent developments in DRL have made it possible to scale to large dimensional state space:

- The combination of duelling DQN architecture with prioritized experience replay in providing better estimates of expected return functions [82, 83].
- The use of an experience replay and target network that initially contains weights of the network enacting the policy, but is kept frozen for a large period [83–85].
- Introduction of hierarchical reinforcement learning.
- Improvements in guided policy algorithms.
- Asynchronous advantage actor-critic (A3C) algorithm [86] developed for both single and distributed machine settings. A3C combines the advantages of updates with actor-critic formulation and relies on asynchronous update policy and value networks in parallel.

1.5.2 Review of imitation learning

Imitation learning (IL) aims to mimic human behaviour in a given task by facilitating the teaching of complex tasks with minimal knowledge through demonstration. There are three main classes of ML algorithms for imitation, namely, behaviour cloning, inverse reinforcement learning and generative adversarial learning [87]. Behaviour cloning applies supervised learning by learning a mapping between the input observation and the corresponding actions, provided there is enough data. Generative adversarial imitation is inspired by generative adversarial networks [88]. Typically, an agent uses instances of performed action to learn a policy that solves a given task using ML techniques. The agent could learn from trial and error or observe other agents. It has been applied to problems in real-time perception and reaction, such as humanoid robots, self-driving cars, human-computer interfaces and computer games. The assumption is that an expert (teacher) is more efficient than the agent learning from scratch when given a task [89]. Imitation learning is an interdisciplinary field of research, and it is sometimes difficult to define suitable reward function for complex tasks. For example, it is often the case that direct imitation of an expert's motion does not suffice due variations in the task such as the position of the object, environmental conditions and inadequate demonstrations [90]. Therefore, it is difficult to learn policies given demonstrations that generalized to unseen scenarios. The policy must be able to adapt to variations in the task and surrounding environment. Argall et al [91] address different challenges in the process of IL, such as computational methods used to learn from demonstrated behaviour and the processing pipeline. A typical representation of a sample for IL consists of pairs of action and state, such as position, velocity and geometric information, and modelling the process as MDP. The learning process is with pre-processing, sample creation and direct or indirect imitation.

The following are some of the challenges of IL [90]: Noisy or unreliable sensing, correspondence problem and observability where the kinematics of the teacher is not unknown to the learner. Further, complex behaviour is often viewed as a trajectory of dependent micro-actions, which violates independent and identically distributed assumptions in machine learning. Lastly, safety concerns in human-robot interactions, the ability of the robot to react to human force and adapt to the task. A typical flow chart [90] is shown in **Figure 4**.

There are different methods from demonstrations, namely, structured predictions [92], dynamic movement primitives [93], inverse optimal control (inverse reinforcement learning [94], active learning [95], transfer learning and other techniques.

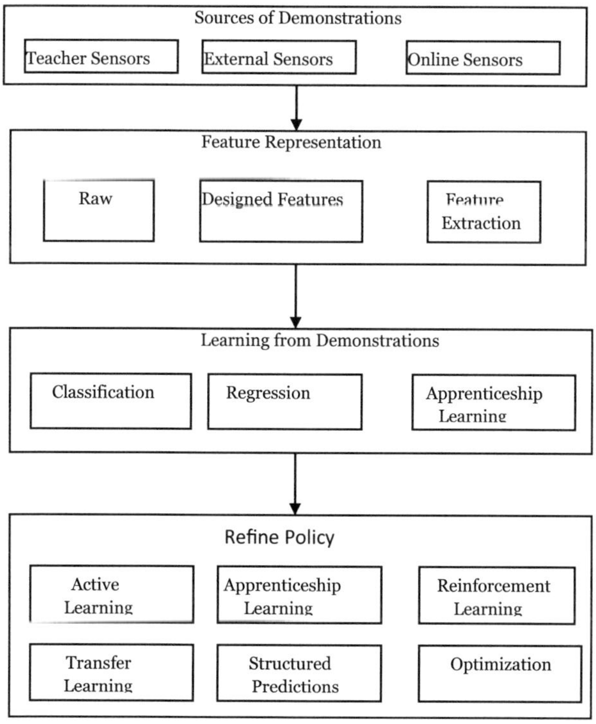

Figure 4.
Imitation learning flowchart [89].

Active learning needs a dedicated oracle that can be queried for demonstration. Inverse RL techniques use demonstrations to learn cost functions over extracted features. It first recovers a utility function that makes the demonstration near-optimal and searches for the optimal policy using a cost function as an optimization objective. Closely related is apprenticeship learning, which uses demonstrations from an expert or observation to learn a reward function. A policy that optimises the reward function is then learned through experience (trial and error). Transfer learning use experience from old tasks or knowledge from other agents to learn a new policy. The reader is referred to refs. [87, 96] for details of imitation learning and its applications in robotics. Learning a direct mapping between state and action is not enough to achieve the required behaviour in most cases due to cascade errors, insufficient demonstrations and the difficulty in reproducing the conditions and settings. The learner has to learn actions and re-optimise policies with respect to quantifiable reward functions. **Figure 4** is a flowchart showing different variants of imitation learning. The following are some recent developments:

- Use of goal-directed (motivation or curiosity-driven) learning to exploit and explore multi-task spaces;

- Use of developmental robotics concept of goal babbling for visuomotor coordination tasks for coordination of multiple subsystems (head and arms) [97].

- Use of predictive processing techniques [3].

- Use of memory systems for storage of knowledge of agents' beliefs, goals and short and long-term memory, together with efficient integration with other components of cognitive architectures.

- Use of machine learning for integration of perceptual processing, feature extraction, learning and control.

1.5.3 Review of deep learning algorithms

The recent success of deep neural networks (DNN) in computer vision and natural language processing has led to its application in cognitive robotics. Traditionally, cognitive robotics architecture has been built with artificial intelligence at the top level using a restricted form of natural language and gestures for communication, and biologically inspired mechanisms at the lower levels. Deep learning using DNN has been applied to perceptual processing, motor control, object manipulation and different cognitive processing level of the generic architecture discussed earlier on. A deep learning survey focusing on deep reinforcement learning and imitation has been provided by Tai et al [81], including applications in ML in robotics. Perception processing is passive since an intelligent agent receives observations from the environment and then infers the desired properties from the sensory input. Guo et al [98] provide a comprehensive overview of deep learning for perception. Similarly, for manipulation applications, Gu et al. [99] present on deep reinforcement learning for robotic manipulations. Gupta et al. [100] also present on robotic manipulations using human demonstrations. Several works relating to deep reinforcement learning in robotic navigation [101–103] have been published including those using SLAM [104, 105]. Zhang et al. [104] propose neural SLAM based on a neural map proposed by Parisotto and Salakhutdinov [106], which in turn uses a neural turing machine for the deep RL agent to interact with. The main challenge with DRL is the reality gap, which refers to discrepancies between models trained with data from simulated environment, transferred to the real world, and deployed on real robotic platforms. It is due to unrealistic environmental conditions such as lighting conditions, noise patterns, texture, etc., synthetic rendering and real-world sensory readings. It is particularly several with visual data (images and videos). Domain adaptations are typical to use to mitigate the problem [107] based on generative adversarial networks (GANS).

Other DNN architectures include convolutional autoencoders for low-dimensional image representation [108], deep recurrent neural networks [109] and deep convolutional networks [11, 110]. To improve robustness of deep learning networks, several strategies have been adapted, including the following: Use of auxiliary tasks in either supervised or unsupervised fashion; experience replay, hindsight experience, curriculum learning, curiosity-driven exploration, self-replay and noise in parameter space for exploration. **Table 1** provides a summary of representative research works covering different ML approaches to solving cognitive problems and the functionality provided. For industrial 4.0, initiative typical ML algorithms are provided in ref. [56].

1.6 Use case

For robots acting as human companions, autonomy is fully-oriented towards navigation in a human-centred environment and human-robot interactions. It is facilitated if the robot's behaviour is as natural as possible. Some requirements are that robot independent movement must appear familiar and predictable to humans and have similar appearance to humans. Human–robot interactions include the following: use of natural language or subset for communication, gesture or activity interpretation that involves tracking and action recognition; gesture imitation that involves tracking and reproduction and the person following which involves 2-D or 3-D based tracking. Acceptable performance at the task level requires real-time processing constraint of 50 milliseconds per second. Safety is also very important as robots are expected to evolve in a dynamic environment, well populate with humans. The main challenge is that robotic systems lack learning representation, and interactions are often limited to

Machine learning paradigm	Reinforcement learning	Imitation learning	Deep learning	End-to-end processing task
Transfer learning	[111] [112] DQN			Games Games
Representational learning			[113] [114]	Object recognition Navigation
Feature extraction			[115] k-means [116] autoencoder [117] autoencoder [118, 119] recurrent neural networks [120] LSTM [121, 122] CNN	Language and behaviour learning Trajectory planning Object grasping
ML plus other techniques	[123]	[124, 125]		

Table 1.
Comparison of different ML techniques reported in the literature.

pre-programmed actions. One solution strategy is to conceptualize cognitive robots as permanent learners, who evolve and grow their capacities in close interactions with users [86]. Robots must learn new tasks and actions relative to humans by observing and imitating (imitation learning). Thus human detection and tracking, activity recognition and face detection are some basic tasks that must be performed robustly in real-time. A use case is presented next, which deals with daily activity recognition at home and face recognition using publicly available dataset. These typically fit in several robotic studies investigated in human-centred environments [40]. The algorithms are first described, followed by an evaluation.

1.6.1 Activity recognition

Research activities in domestic service robots have increased in recent years. Some of the main drivers are the projected future use of domestic robots for improving elderly people's quality of life, childcare, entertainment and education. Several benchmark datasets [126–129] and methodologies for evaluating the capabilities and performance of robotic platforms are available. Action recognition is used in several application domains such as surveillance, patient monitoring systems, human–computer interface, housekeeping activities and human assistance by robots (guiding humans). There are two processing techniques: spatial approach, which allows recognizing activities from images, and spatio-temporal approach for detecting specific activity as space-time volume.

The HMDB51 [130] is an action dataset whose action categories mainly differ in motion rather than static points. It contains 51 distinct action categories, each containing at least 101 video clips. Video clips are extracted from a wide range of sources. The clips have been annotated and validated by at least two human observers. Additionally, meta information tags allow for a precise selection of tags for training, testing and validation. Meta-data tags include information on camera viewpoint, presence or absence of camera motion, video quality and a number of actors involved. The training procedure is also described.

A simulation study on activity recognition based on spatio-temporal analysis of a large video database of human motion recognition [130] is provided. The main processing steps are shown in **Figure 5**. The algorithm consists of six main processing steps, namely, pre-processing, spatio-temporal analysis in the wavelet domain, class model construction (class dictionary), batch singular value factorization (BSVF), similarity feature computation and classification. The pre-processing step involves filtering for noise removal and optionally contrast enhancement using histogram equalization.

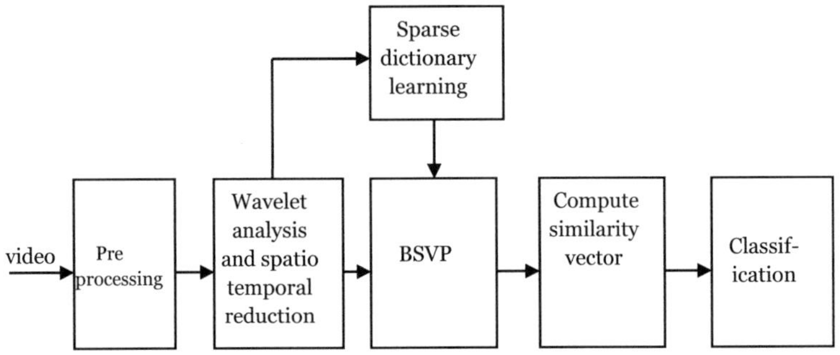

Figure 5.
Similarity-based feature construction and classification.

The wavelet analysis step applies orthogonal or biorthogonal wavelet (9/7 or 5/3 filter) to produce subband frames. A silhouette feature map is constructed by combining low–low and high–low subbands as described in Tawiah et al [130]. The map is a tiling of rectangular features describing the dominant objects in the frame. Sparse dictionary is constructed for each activity as described in refs. [131, 132]. Spatial frame resizing and temporal frame subsampling by interpolation are applied to construct an action volume of $64 \times 32 \times 100$ pixels for each action volume. It is then reshaped to a vector of size of 51200. Batch singular value prediction (BSVP) is based on the classical singular value decomposition [133] used in signal processing with batch data input (matrix). Each column of the input matrix represents a sample action. The output is a decomposition consisting of left and right-hand singular vectors (or matrices) for vector (or matrix) input and a covariance matrix as the diagonal matrix.

BSVP prediction step consists of two sub-steps: first, apply singular value decomposition to the same batch training sample used in constructing the dictionary, replacing one column (e.g, the first) with an incoming action sample. Then, apply the computation step in Eq. (5). The class dictionary is constructed using a batch sample matrix, with each sample representing an action volume. The prediction for an input action sample is computed using Eq. (5):

$$Est(r,j) = \sum_{j=1}^{nsample} \varphi(r,\ :) \left[\sum_{j=1}^{\text{dims}} LHS(r,\ i) * \alpha(j,\ j) + \sum_{j=1}^{\text{dims}} RHS(r,\ j) * \alpha(j,\ j) \right] \tag{5}$$

Φ denotes the class dictionary matrix, N sample denotes the number of samples in the batch dataset, Dim S denotes the dimension of each sample, RHS (r,i) denotes the right-hand singular vector, LHS(r,i) denotes the left-hand singular vector, α denotes the covariance matrix and Est denotes the estimate of the sample. The indices, r and i, are used to identify specific elements in a matrix. The similarity between the input spatio-temporal volume and Est (refer to Eq. 5) is computed using five similarity measures, namely, canonical correlation [134], Bhattacharyya distance [135], modified Bhattacharyya distance, histogram intersection [136] and cityblock. A similarity vector is formed by concatenating all the similarity values. A multi-class feed-forward classifier [137], consisting of 51 all versus one classifier, is constructed. The classifier is

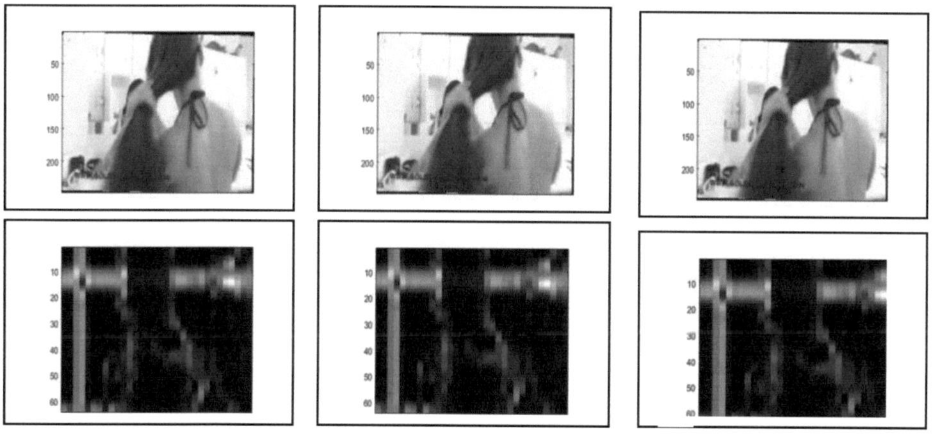

Figure 6.
Brush hair sample video clip, showing frames 1, 2 and 3.

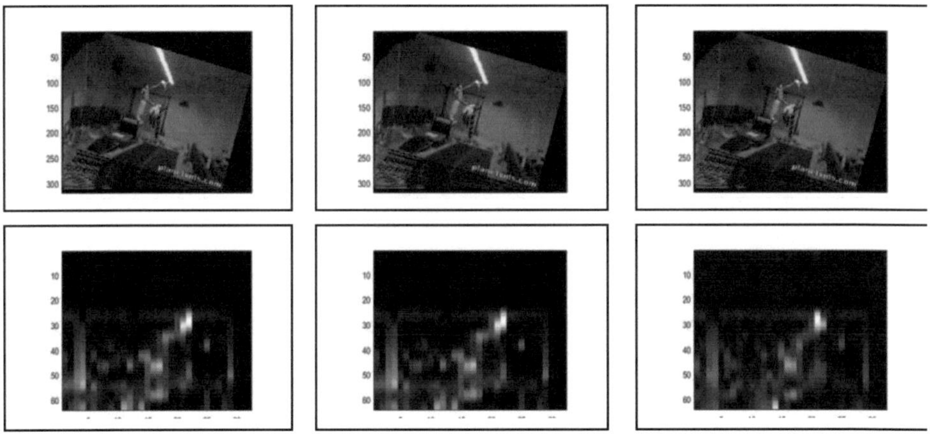

Figure 7.
Cartwheel sample video clip, showing frames 1, 3 and 5.

able to assign an action volume to multiple classes. Samples of input video frames and the corresponding object outline maps are shown in **Figures 6** and 7.

BSVP does not reconstruct a sample using the sparsest representation as is the case in classical sparse coding but instead uses one-time reconstruction from batch sample whose representations are known (represented as LHS and RHS singular matrices with known covariance) and applies BSVP algorithm. This provides a representation for a sample taking into consideration statistical characteristics of all samples in the batch. It is computationally efficient and avoids solving L1-norm optimization, and it is suitable for real-time classification problems. The result on applying the proposed algorithm to all the fifty-one action classes is summarised in **Figure 8**, using the action categories provided by HMDB51 dataset (**Table 2**).

The confusion matrix is also shown in **Figure 8** to illustrate action classes prone to misclassification.

For robotics applications, facial expression recognition and gesture recognitions are also very important. Reference [138] provides a good review of facial expression recognition.

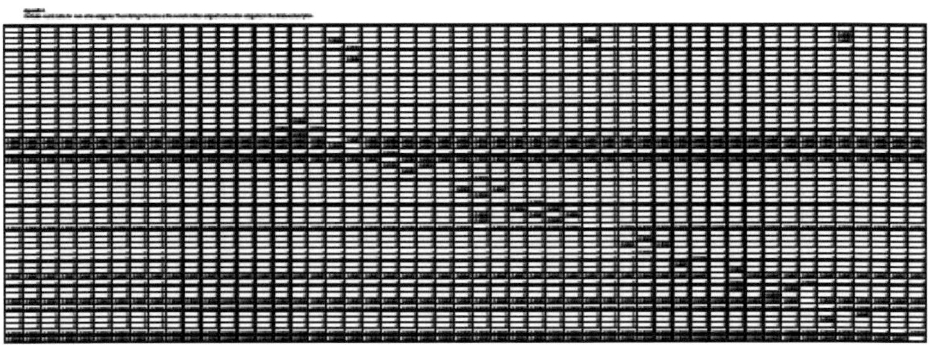

Figure 8.
Confusion matrix for HMDB51 dataset.

Action class	Scene content precision (%)	Meta descriptors precision (%)
Facial expression	75	25
Facial action	78	23
Body movement	86	37
Body movement with object interaction	97	43
Body movement with human interaction	97	43

Table 2.
HMDB 51 action classification.

1.7 Trends in cognitive robotics

Early approaches to imitation aim to reproduce reaching or grasping with simple grippers. Imitation learning provides a desired sequencing of basic sub-skills to achieve an observed task behaviour. Later on, more sophisticated system including modules for visual attention, speech recognition, and integration of visual and linguistic inputs for instructing robots to grasp everyday objects [139]. Online learning and machine learning techniques, such neural networks, have been used in low-level and reactive tasks from trajectory learning and adaptive control of multi-DOF robots, and tasks learning from demonstrations. ML provides different paradigms of learning from transfer learning, representation learning curriculum learning, etc, which provides for systematic means of acquiring systematic models for making inferences [140]. The following are some trends that are apparent from the literature review:

- The use of neuroscience and behavioural psychology to synthesize computational models for high-order cognitive skills in artificial agents.
- The use of neural networks as functional approximators.
- Use of motivations or goal-directed mechanisms to balance exploration and exploitation in tasks space rather than in motor space.
- Use of robotic platforms for research in higher order research (social robots).
- Use of predictive coding mechanisms to synthesise higher-order cognitive behaviours.
- Classical control theory is unable to handle complex scenarios with many parameters.
- Use of swarm robotics to study social behaviours in robotic swarms.
- The increasing use of networked and cloud robotics and cyber-physical systems.

The use of artificial intelligence, especially machine learning, wireless connectivity and cloud computing, is increasing to integrate physical systems and processes, including robotics. At the core of most ML tasks, decision-making is based on information fed to the decision maker. The study of decision-making is closely connected with psychology and cognitive sciences.

1.8 Success, challenges and research directions

Several projects involving the use of cognitive robotics have been reported in industrial settings (Industry 4.0), service robots, robotic surgery, cardiovascular surgery [70], assistive technology [141] and several other fields. In ref. 70, ML methods are used in perceiving an individual's health by collecting and interpreting his/her clinical data and would reason to suggest actions to maintain or improve the individual's cardiovascular health. ML augmented decisions point to potential to improve the outcome at a lower cost of care and increase satisfaction. In assistive technology [141], vision-based hand wheelchair control using kinect sensor system enables the user to control without wearing or touching.

As cognitive robotics continues to make some remarkable progress in industrial process automation with Industry 4.0 initiative, cloud robotics and service robots, it has resulted in more challenges [142–144]. For example, standardisation effort [56] has ushered in a new era of robotics linked to cyber–physical system for effective control and monitoring of industrial processes. Classical approaches to robotics have made significant progress in control-based applications in stand-alone robotic applications, but there are challenges in multi-robot and multi-agent systems applied to complex tasks in dynamic environments.

The main goal of integrating thinking, knowing and feeling in an artificial intelligent system as cognitive process has not been realised today despite advances [57]. In particular, integrating feeling into the existing system has proved very challenging.

The trends towards Industry 4.0 of providing cyber–physical framework for unifying industrial processes and cyber–physical system would be extended to service robots domain as well. The need to develop more robust and sophisticated ML algorithms to enable AI agents to carry out complex tasks in a coordinated and cooperative fashion to ensure reliability and cost-effectiveness. The robustness of ML algorithms under adversarial learning would also have to be investigated.

The need is for more research into the decision-making process (using ML) to make it robust, timely and relevant to situation, as well as meet real-time requirements. For multi-robot systems, the need for cooperation and coordination of tasks is very challenging to improve the effectiveness and improved utilisation of resources. Underlying these problems is the need for research into more robust ML algorithms and transparent model interpretation, and guarantees against adversarial attacks [145].

The need is for cost-effective management of resources (computing, network, storage and devices), all interconnected for ambient intelligence. The problem of scheduling, recovering from unexpected events and scalability issues require urgent attention. Similarly, the integration of heterogeneous platforms (software and hardware) into processes is required. Investigations into robust and generic processing architecture for social robotics are another area worthy of investigation.

Investigations into protocols to ensure effective and robust cooperation between humans and robots via human-machine interfaces to ensure trust and autonomy, as well as ethical considerations, ought to be investigated.

To meet privacy and security concerns distributed learning [146, 147] approaches to train models on the cloud keeping data localized and apply privacy-preserving analysis. However, this has raised the issues of network latency and model consistencies, which has been proven very challenging. Approaches to solving the challenges include MEC-based training, federated learning and capsule network for internet of vehicles. Other persistent challenges are latency, security and management of network infrastructure. Autonomic systems [148] seem very attractive for managing problems related to network and computing infrastructure.

1.9 Conclusion

The chapter has presented a review of recent development in ML techniques for cognitive robotic systems in the overall context of artificial intelligence. The main algorithms for learning, namely, reinforcement and imitation learning techniques, have been discussed.

The recent initiative in Industry 4.0 initiative, increasing trend in research in service robots, telemedicine and computer-assisted medical delivery system means that the industry holds lots of promise for research and personal applications.

Several processing architectures, as well as software frameworks for integrating heterogeneous hardware and software components, have also been presented. Towards simulating further research, current trends and research issues have also been highlighted. An example scenario involving action recognition of humans and facial expression has also been presented.

Author details

Thomas Tawiah
University of Education, Winneba, Ghana

*Address all correspondence to: thomastawiah@yahoo.co.uk

References

[1] Hacker M. Humanoid Robots: Human-like Machines. Vienna, Austria; 2007. pp. 367-396

[2] Jurgen J. A bottom-up integration of vision and actions to create cognitive humanoids. In: Samani H, editor. Cognitive Robotics. Boca Raton, FL: CRC; 2015. pp. 191-214

[3] Ciria A, Schillaci G, Pezzulo G, Hafner VV, Lara B. Predictive processing in cognitive Robotics: A review, 2021

[4] Kambayashi Y, Yajima H, Shiyoji T, Oikawa R, Takimoto M. Formation control of swarm robots using mobile agents. Vietnam Journal of Computer Science 2019;6(2):193-222

[5] Schillaci G, Hafner V, Lara B. Exploration behaviors, body representations, and simulation processes for the development of cognition in artificial agents. Frontiers in Robotics and AI;**3**:39

[6] Alami R, Chatila R, Fleury S, Ghallab M, Ingrand F. An architecture for autonomy. International Journal of Robotics Research (Special Issue on Integrated Architecture for Robot Control and Programming). 1998;**17**: 315-337

[7] Sun B, Saenko K. From virtual to reality: Fast adaptation of virtual object detectors to real domains. BMVC. 2014;**1**

[8] Tobin et al. Domain randomization for transferring deep neural networks from simulation to the real world, March 2017

[9] Tzeng E, Hoffman J, Zhang N, Saenko K, Darrell T. Deep domain confusion. Maximizing for domain invariance. 2014

[10] Tzeng E, Devin C, Hoffman J. et al. Adpating deep visuomotor representations with pairwise constraints. 2017

[11] Levine S, Finn C, Darrell T, Abbeel P. End-to-end training of deep visuomotor policies. Journal of Machine Learning Research. 2016;**17**:1-40

[12] Guido S. Sensorimotor Learning and Simulation of Experience as a Basis for the Development of Cognition in Robotics. Germany: Humboldt University of Berlin; 2013

[13] Hayamizu S, Hasegawa O, Itou K, Yoshimura T, Akiba T, Asoh H, Kurita T, Sakaue K. Multimodal interaction systems that integrates speech and visual information. Bulletin of the Electrotechnical Laboratory 2000;64(4-5):37-44

[14] Steels L, Kaplan F. Aibos first words. The social learning of language and meaning. Evolution Communication. 2001;**4**(1):3-21

[15] Iwahashi N. Language acquisition through a human-robot interface by combining speech, visual, and behaviour information. Information Science. 2003; **156**:109-121

[16] Brooks RA. A robust layered control system for a mobile robot. IEEE Journal of Robotics and Automation. 1986;**RA-2**:1

[17] Agostini A, Torra C, Worgotter F. Efficient interactive decision-making framework for robotic applications. Artificial Intelligence 2017;247:187-212

[18] Yeon ASA, Visvanathan R, Mamdah SM, Kamarudin K, Kamarusin LM,

Zakaria A. Implementation of behavior based robot with sense of smell and sight. In: 2015 IEEE International Symposium on Robotics and Intelligent Sensors (IRIS 2015). 2015. pp. 119-125

[19] Zucker M, Ratliff N, Stolle M, et al. Optimization and learning for rough terrain legged locomotion. The International Journal of Robotics Research;**30**(2):175-191

[20] Schillaci G, Ciria A, Lara B. Tracking emotions: Intrinsic motivation grounded on multi-level prediction error dynamics. In: Proceedings of the 10th Joint International Conference on Development and Learning and Epigenetic Robotics (IEEE ICDL-EpiRob 2020). 2020

[21] Pio-Lopez L, Ange N, Fristorn K, Pezzulo G. Active inference and robot control: A case study. Journal of Royal Society Interface. 2016;**12**:616

[22] Buckley C, Kim CS, Mcgregor S, Seth AK. The free energy principle for action and perception: A mathematical review. Journal of Mathematical Psychology;**81**: 55-79

[23] Lara B, Astorga D, Mendoza-Bock E, Pardo M, Escobar E, Ciria A. Embedded Cognitive robotics and the learning of sensorimotor schemes. Adaptive Behaviour;**26**(5):225-238

[24] Pickering M, Clark A. Getting ahead: Forward models and their place in cognitive architecture. Trends in Cognitive Sciences;**18**(9):451-454

[25] Lanillos P, Cheng G. Adpative robot body learning and estimation through predictive coding. In: Proceedings 2018 IEEE/RSJ International Conference on Intelligent Robotics and Systems (IROS). 2018. pp. 4083-4090

[26] Lanillos P, Cheng G et al. Robot self/other distinction: Active inference meets neural networks in a mirror. 2004

[27] Asada M, Hosoad K, Kuniyoshi Y, et al. Cognitive developmental robotics: A survey. IEEE Transactions on Autonomous Mental Development. 2009;**1**(1):12-34

[28] Pfeifer R, Lida F, Gomez G. Morphological computation for adaptive behaviour and cognition. International Congress Series. 2006;**1291**:22-29

[29] Mcgeer T. Passie walking with knees. In: Proc. 1990 IEEE Int. Conf. Robot Autom. 1990

[30] Sumioka H, Yoshikawa Y, Asada M. Development of joint attention related actions based on reproducing contingency. In: Proceedings of 7th International Conference on Developmental Learning. 2008

[31] Hashimoto T, Senda M, Kobayashi H. Realization of realistic and rich facial expressions by face robot. In: Proceedings of 2004 IEEE Techn. Exhib. Based Conf. Robot Autom. 2004. pp. 37-38

[32] Matsui D, Minato T, MacDorman KF, Ishiguro H. Generating natural motion in an android by mapping human motion. In: Proceedings IEEE/RSJ Int. Conf. Intell. Robots Sys. 2005. pp. 1089-1096

[33] Tobin J, Fang A, Scheider R, Zaremba W, Abbeel P. Domain randomization for transferring deep neural networks from simulation to real world. 2017

[34] Kawamua K, Brown W. Cognitive robotics' Chapter in Springer Encyclopedia of Complexity and System

Science. Springer Science; 2010. pp. 1109-1126

[35] Melta G, Fitzpatrick P, Natale T. YARP: Yet Another Robot Platform. International Journal of Advanced Robotic Systems, Special Issue on Software Development and Integration in Robotics. 2006;**3**(1)

[36] Frank M, Leitner J, Stollenga M, Harding S, Forster A, Schmidhuber J. The modular behavioural envirnment for humanoids and other robots (MoBeE). In: Proceedings of the International Conference on Informatics in Control, Automation & Robotics (ICINCO). 2012

[37] Stollenga M, Pape L, Frank M, Leitner J, Forster A, Schmidhuber J. Task-relevant roadmaps: A framework for humanoid motion planning. In: Proceedings of the International Conference on Intelligent Robotics and Systems (IROS). 2013

[38] Leitner et al. A modular software framework for hand-eye coordination in humanoid robots. Frontiers in Robitics and AI. 2016;**2016**:1-16

[39] Courtney et al Cognitive systems platforms using open source. 2009

[40] Correa M, Hemosilla G, Verschae R, Ruiz-del-solar J. Human detection and identification by robots using therma and visual information in domestic environments. Journal of Intelligent Robotic Systems;**66**:223-243

[41] Cheraghi AR, Shahzad S, Graffi K. Past, present, and future of swarm robotics. 2021

[42] Baranes A, Oudeyer P-Y. Intrinsically motivated goal exploration for active motor learning in robots: A case study. In: Proc: IEEE/RSJ International Conference on Intelligent Robots and Systems. 2010. pp. 1766-1173. DOI: 10.1109/IROS.2010.5651385

[43] Sun R. The importance of cognitive architecture: An analysis based on CLARION. Journal of Experimental and Theoretical Artificial Intelligence. 2007; **19**(2):159-193

[44] Sun R. Anatomy of the Mind. Oxford University Press; 2016

[45] John E. The Soar Cognitive Architecture. MIT Press; 2012. p. 390

[46] Demiris Y. Predicition of Intent in Robotics and Multi-agent systems. Cognitive Processing. 2007;**8**:151-158

[47] Demiris Y, Khadhouri B. Hierarchical attentive multiple models for execution and recognition of actions. Robotics and Autonomous Systems. 2006;**54**:361-369

[48] Vahrenkamp N, Wachter M, Krohnert M, Welke K, Asfour T. The robot software framework Armarx. Information Technology. 2015;**57**(2): 99-111

[49] Meta G et al. The iCub humanoid robot: An open-systems platform for research in cognitive development. Neural Networks. 2010

[50] Unity Technologies. 2019 Available: https://unity.com

[51] Juliani A, Berges V-P, Teng E, et al. Unity: A general platform for intelligent agents. 2020

[52] Bellemare MG, Naddaf Y, Veness J, Bowling M. The arcade learning environment: An evaluation platform for general agents. In: Proc. International Joint. Conference on Artificial Intelligence. 2015. pp. 253-279

[53] Mnih V, Badia AP, Mirza M, Graves A, Lillicrap TP, et al. Asynchronous methods for deep reinforcement learning. In: Proc. Int. Conf. Learning Representation. 2016

[54] Sutton RS, Barto AG. Reinforcement Learning: An Introduction. Cambridge, MA: MIT Press; 1998

[55] Kehoe et al. A survey of research on cloud robotics and automation. IEEE Transactions on Automation Science and Engineering. 2015;**12**(2):398-409

[56] Groshev M et al. Toward Intellignt Cyber-Physical Systems: Digital twin meets artificial Intelligence. IEEE Communications Magazine. 2021;**59**(8): 14-20

[57] Gutierrez-Garcia J. and O. Lopez-Neri. Cognitive computing: A brief survey and open research challenges. In Proceedings of 2015 3rd International Conference on Applied Computing and Information Technology/2nd International Conference on computational Science and Intelligence, Japan, 2015.

[58] Brasil L et al. Hybrid expert systems for decision support in the medical area: Complexity and cognitive computing. International Journal of Medical Informatics. 2001;**63**(11):19-30

[59] Wang Y. Towards the synergy of cognitive informatics, neural informatics, brain informatics, and cognitive computing. In: Cognitive Information for Revealing Human Cognition: Knowledge Manipulations in Natural Intelligence. First ed. Hershe, PA, USA: IGI Global; 2012. pp. 159-177

[60] Cote et al. Prototyping cognitive models with MARIE. In: Proceedings on (IEEE/RSJ 2008 International Conference on Intelligent Robots and Systems), IROS Workshop on Current Software Frameworks in Cognitive Robotics Integrating Different Computational Paradigms. Nice, France; 2008

[61] Cote C, Letournrau D, Raievsky C, Michaud F. Robotic software integration using MARIE. International Journal of Advanced Robotic Systems. 2006;**3**(1): 55-60

[62] Robot operating system (ROS) https://en.wikipedia.org/wiki/Robot_Operating_System

[63] ROS 2 for Realtime applications. https://discourse.ros.org/t/ros2-for-real-time-applications/6493. ROS.org. Open Robotics, 17 October 2018

[64] Nao ROS Wiki. http://www.ros.org/wiki/nao ROS.org. Open Robotics, 28 October 2013

[65] Pereira A, Bastos GS. ROSRemote, using ROS on cloud to access robots remotely. In: Proceedings of the 2017 IEEE 18th International Conference on Advanced Robotics (ICAR). Hongkong, China; 2017

[66] Arumugan R et al. Davinci: A cloud computing framework for service robots. In: Proceedings of the 2010 IEEE International Conference on Robotics and Automation (ICRA). Anchorage, AK, USA; 2010. pp. 3084-3089

[67] Multi-Access Edge Computing (MEC): Framework and reference architecture; https://www.etsi.org/deliver/etsi gs/MEC/001099/003/02.02.0160/gsmec003v020201p.pdf

[68] Borsatti et al. Enabling industrial IOT as a service with multi-access edge computing. IEEE Communication Magazine. 2021;**59**(8):21-27

[69] Kahneman D. Thinking Fast and Slow. first ed. Farrar, Straus and Giroux; 2011

[70] Sanchez-Martinez M et al. Machine learning for clinical decision making: Challenges and opportunities in cardiovascular imaging. Frontiers in Cardiovascular Medicine. 2022

[71] Aoki et al. Human-Robot cooperation for autonomous vehicles and human drivers: Challenges and solutions. IEEE Communications Magazine. 2022;**59**(8):36-41

[72] Schillaci G, Villapando AP, Hafner VV, Hanaper P, Colliaux D, Wintz T. Intrinsic motivation and episodic memories for robots exploration of high-dimensional sensory spaces. Adaptive Behaviour. 2020;**29**(6):549-566

[73] Baranes A, Oudeyer P. R-IAC: Robust intrinsically motivated exploration and active learning. IEEE Transactions on Autonomous Mental Development, IEEE. 2009;**1**(3): 155-169. DOI: 10.1109/ TAMD.2009.2037513

[74] Lee K, Ognibene D, Chang HJ, Kim TK, Demiris Y. STARE: Spatio-Temporal attention relocation for multiple structured activities detection. IEEE Transactions on image processing. 2015; **24**(12):5916-5927

[75] Demiris Y. Prediction of intent in robotics and multi-agent systems. Cognitive Processing. 2007;**8**:151-158

[76] Mcleland J, McNaughton L. Why there are complementary learning systems in the hippocampus and Neocortex: Insights from the success and failures of connectionist models of learning and memory. Psychological Review. 1995;**102**(3):419-457

[77] Bellman R. On the theory of dynamic programming. Proceedings of the National Academy Science. 1952;**38**(8): 716-719

[78] Arulkumaran K, Peter M, et al. Deep reinforcement learning: A brief survey. IEEE Signal Processing. 2017; **34**(6):28-38

[79] Bilard A, Calinan S, Dillman RR, et al. Robot Programming by Demonstration. Springer. pp. 1371-1391

[80] Tai L, Zhang J, Liu M et al. A survey of deep network solutions for learning control in robotics: From reinforcement learning to imitation. 2016

[81] Schad T, Quan J, Antonoglou I, Silver D. Prioritized experience replay. In: Proc. Int. Conf. Learning Representations. 2016

[82] Hasselt HV. Double Q-learning. In: Proc. Neural Information Processing Systems. 2010. pp. 2613-2621

[83] Lillicrap TP, Hunt JJ, Pritzel A, et al. Continuous control with deep reinforcement learning. In: Proc. Int. Conf. Learning Representations. 2016

[84] Mnih V, Silver D, et al. Human-level control through deep reinforcement learning. Nature. 2015;**518**(7540): 529-533

[85] Kaelbling LP, Littman ML, Cassandra AR. Planning and acting in partially observable stochastic domains. Artificial Intelligence. 1998;**101**(1): 99-134

[86] Nachum O, Nourouzi M, Xu K, Schuurmans D. Bridging the gap between value and policy based reinforcement learning

[87] Ho J, Ermon S. Generative adversarial imitation learning. 2016

[88] Schmerling M, Schillaci G, Hafner V. Goal-directed learning of hand-eye coordination in a humanoid robot. In: Proceeding of 5th International Conference on Development and Learning and on Epigenetic Robotics (ICDL-EpiRob). 2015

[89] Hussein A, Gaber MM, Elgan E, Jayne C. Imitation learning: A survey of learning methods. ACM Computing Surveys. 2017;**50**(2):35

[90] Argall B, Browning B, Veloso M. Learning by demonstration with critique from a human teacher. In: Proceedings of the ACM/IEEE International Conference on Human-Robot Interaction, SCM. 2007. pp. 57-64

[91] Bitzer S, Vijayakumar S. Latent spaces for dynamic movement primitives. In: Proc. 9th IEEE-RAS International Conference on Humanoid Robots (Humanoids '09). 2009

[92] Bagnell JA. An invitation to mitiation. Pittsbrgh, PA: Carnegie-Mellon University; 2015

[93] Abbeel P and Ng A. Y. Apprenticeship learning via inverse reinforcement learning. In: Proceedings of the 21st International Conference on Machine Learning (ICML), 2004

[94] Baraness A, Ouder P-Y. Active learning of inverse models with intrinsically motivated goal exploration in robots. 2013

[95] Daume H, Langford J, Marco D. Search-based structured prediction. Machine Learning. 2009;**75**:297

[96] Duan Y, et al. One-shot imitation learning. In: Advances in Neural Information Processing Systems. 2017. pp. 1087-1098

[97] Guo Y, Liu Y, Oerlemans A, Laos S, Wu S, Lew MS. Deep learning for visual understanding: A review. NeuroComputing. 2016;**187**:27-48

[98] Gu S, Holy E, Lillicrap T, Levine S. Deep reinforcement learning for robotic manipulation with asynchronous off policy updates. In: Proceedings of IEEE International Conference on Robotics and Automation (ICRA). 2017. pp. 3386-3396

[99] Gupta A, Eppner C, Levine S, Abbeel P. Learning dexterous manipulations for a soft robotic hand from human demonstrations. In: Proceedings of IEEE/RSI International Conference on Intelligent Robots and Systems (IROS). 2016. pp. 3786-3793

[100] Zhang J, Springenberg JT, Boedecker J, Burgard W. Deep reinforcement learning with successor features for navigation across similar environments. In: Proceedings of IEEE/RSJ International Conference on Intelligent Robots and Systems (IROS). 2017. pp. 2371-2378

[101] Chen Y, Everett M, Liu M, How JP. Socially aware motion planning with deep reinforcement learning

[102] Tai L, Paolo G, Liu M. Virtual-to-real deep reinforcement learning: Continuous control of mobile robots for maples navigation. In: Proceedings of IEEE?RSJ International Conference on Intelligent Robots and Systems (IROS). 2017. pp. 31-36

[103] Zhang J, Tai L, Boedecker J, Burgard W, Liu M. Neural SLAM

[104] Khan A, Zhang C, Atanasov N, Karydis K, Kumar V, and Lee D. Memory augmented control networks

[105] Parisotto E, Salakhutdinov R. Neural map: Structured memory for deep reinforcement learning

[106] Tzeng F, Hoffman J, Zhang N, Saenko K, and Darrel T. Deep domain confusion: Maximizing for domain invariance

[107] Chen L-C, Papandreou G, Kokkinos I, et al. Semantic image segmentation with deep convolutional nets and fully connected crfs. In: Proceedings of International Conference on Learning Representations (ICLR). 2015

[108] Fritzpatrick P, Metta G, Natale L, Rao S. Learning about objects through action-initial steps towards artificial cognition. In: Proceedings of International Conference on Robbotics and Automation (ICRA '03). Taipei, Taiwan. pp. 3140-3145

[109] Masci J, Meier U, Cirecsan D, Schmidhuber J. Stacked convolutional auto-encoders for hierarchical feature extraction. In: Proceedings of International Conference on Artificial Neural Networked. Springer; 2011. pp. 52-59

[110] Kanitscheider I, Fiete I. Training recurrent networks to generate hypothesis about how the brain solves hard navigation problems. Advances in Neural Information Processing Systems: 4532-4541

[111] Taylor M, Stone P. Cross-domain transfer for reinforcement learning. In: Proc. 24th International Conference on Machine Learning (ICML'07). 2007. pp. 879-886

[112] Wang et al. Dueling network architectures for Deep reinforcement learning. In: Proc. 33rd International Conference on Machine Learning (ICML'16). 2016. pp. 1995-2003

[113] Radford R, Luke M. Unsupervised representation learning with deep convolutional generative adversarial networks.2016

[114] Ran et al. Convolutional neural network-based robot navigation using uncalibrated spherical images. Sensors. 2017;**17**:1341

[115] Coates A, Ng AY. Learning Feature Representations with k-means. Springer; 2012. pp. 561-580

[116] Schillaci G et al. Intrinsic motivation and episodic memories for robot exploration of high-dimensional sensory spaces. 2020

[117] Chen R, Jin Y. A social learning particle swarm optimization algorithm for scalable optimization. Information Science. 2015;**291**:43-60

[118] Rahmatizadeh R, Abolghasemi P, Bolani L. Learning manipulation trajectories using recurrent neural networks. 2016

[119] Yamada T, Murata S, Aric H, Ogata T. Dynamic integration of language and behavior in a recurrent neural network for human-robot interaction. Frontiers in Neurobotics. 2016

[120] Molina-Leal A et al. Trajectory planning for mobile robot in a dynamic environment using LSTN neural network. Applied Science. 2021;**11**(22): 10689

[121] Redmon J, Angelova A. Real-time grasp detection using convolutional neural networks, 2015

[122] Levine S et al. Learning hand-eye coordination for robotic grasping with deep learning and large scale data collection. 2016

[123] Li C, Lowe R, Ziemke T. Humanoids learning to walk: A natural CPG-actor critic architecture. Frontiers in Neurobiotics. 2013

[124] Calinion S, Li S, Alizadeh T, Tsagarakis G. Statistical dynamical systems for skills acquisition in humanoids. In: Proc. 2012 IEEE-RAS International Conference on Humanoids Robots (Humanoids '12). pp. 232-329

[125] Triesch J, Wirghardt J, Mael E. Towards imitation learning of Grasping movements by an autonomous robot. In: Proc. of the Interantional Gesture Workshop on Gesture-Based Communication in Human-Computer Interaction (GW '99). 1999. pp. 73-84

[126] Wisspeinter T, Van der Zant T, Ioccchi I, Schiffers S. Robocupmome: Scientific competition as benchmarking for domestic robots. Interaction Studies. 2009;**930**:392-426

[127] RoboCupHome Official Website. Available on December 2010

[128] Vrigkas M, Christophorous N, Ioannis A. A Review of human activity recognition methods. Frontiers in Robotics and AI. 2015;**2**:28

[129] Kuehne H, Jhuang H, Garrote E, Poggio T, Sierre T. A large video database for human motion recognition. In: Proceedings of the IEEE International Conference on Computer Vision. 2011

[130] Andzi-Quainoo TT, Mike LR. A bank of classifiers for robust object modeling in wavelet domain. In: Proceedings of IEEE International Conference on Industrial Technology. Busan, South Korea; 2014

[131] Lee H, Alexis B, Rajat R, Ng Andrew Y. Efficient sparse coding algorithms. In: Advances in Neural Information Processing (NIPS). 2007

[132] Aharon M, Elad M, Bruckstein AM. K-SVD: An algorithm for designing overcomplete dictionaries for sparse representations. IEEE Transactions on Signal Processing. 2006;**54**(11): 4311-4322

[133] Golub GH, Reinsch C. Singular value decomposition and least squares solutions. Numerical Mathematics. 1970; **14**:403-420

[134] Hardle HW, Leopold S. Applied Multivariate Statistical Analysis. Berline, Heidelberg: Springer; 2007. pp. 321-330

[135] Kailath T. The dirvergence and Bhattacharyya distance measures in signal selection. IEEE Transactions on Communication Technologies. 1967; **15**(1):54-60

[136] Swain M, Ballard DH. Color indexing. International Journal of Computer Vision. 1991;7(1):11-32

[137] Cheng-Liu L. One-versus_all training of prototype classifier for pattern classification and retrieval. In: Proceedings of 2010 20th International Conference on Pattern Recognition. 2010. pp. 3328-3331

[138] Pramerdorf C and Kampel M. Facial expression recognition using convolutional neural networks: State of the art. 2016.

[139] Steil J, Wersing H. Recent trends in online learning for cognitive robotics. In: Proceedings of ESANN '2006 –European Symposium on Artificial Neural networks. Bruges, Belgium; 2006

[140] Tani J. Learning to generate articulated behavior through bottom-up

and top-down interaction processes. Neural Computation. 2003;**16**(1):11-23

[141] Trigueiros P, Ribeiro F. Vision-based hand wheelchair control. In: Proc. of 12th International Conference on Autonomous Robot Systems and Competitions (Robotics 2012). Guimaraes, Portugal; 2012. pp. 39-43

[142] Saha O, Dasgupta P. A comprehensive survey of recent trends in cloud robotics architectures and Applications. Robotics. 2018;7:47

[143] ISO 8373. (en). Robots and robotic devices-vocabulary: ISO/TC299. 2012. https://www.iso.org/obp/ui/#iso:std:iso:8373:ed-2:v1:en [Accessed: [November 2020]

[144] Fong T, Noorbakhsh I, Dautenhahn K. A survey of socially interactive robots. Robotics and Autonomous Systems;**42**:143-166

[145] Modas A, Sanchez-Matilla R, Frossard P, Cavallaro A. Toward robust sensing for autonomous vehicles, an adversarial perspective. IEEE Signal Processing Magazine. 2020;**47**:14-24

[146] Xu et al. Capsule network distributed learning with multi-access edge computing for internet of vehicles. IEEE Communications Magazine. 2021;**59**(8):52-57

[147] Li L et al. A survey on federated learning. In: 2020 IEEE International Conference on Control & Automation (ICCA). 2020. pp. 791-796

[148] Maidana RG et al. Autonomic computing towards resource management in embedded mobile robots. In: Proceedings of 2019 Latin American Robotics Symposium (LARS), 2019 Brazilian Symposium on Robotics (SBR) and 2019 Workshop on Robotics in Education (WRE). 2019. pp. 192-197